BEN JACOBSEN AND
DAVID BEER

SOCIAL MEDIA AND THE AUTOMATIC PRODUCTION OF MEMORY

Classification, Ranking and the
Sorting of the Past

BRISTOL
UNIVERSITY
PRESS

First published in Great Britain in 2021 by

Bristol University Press
University of Bristol
1–9 Old Park Hill
Bristol
BS2 8BB
UK
t: +44 (0)117 954 5940
e: bup-info@bristol.ac.uk

Details of international sales and distribution partners are available at
bristoluniversitypress.co.uk

British Library Cataloguing in Publication Data
A catalogue record for this book is available from the British Library

ISBN 978-1-5292-1815-2 hardcover
ISBN 978-1-5292-1816-9 ePub
ISBN 978-1-5292-1817-6 ePdf

The right of Ben Jacobsen and David Beer to be identified as the authors of this work has been
asserted by them in accordance with the Copyright, Designs and Patents Act 1988.

Cover design: blu inc, Bristol
Cover image: Guille Faingold – stocksy.com

Printed and bound by CPI Group (UK) Ltd, Croydon, CR0 4YY

Contents

Acknowledgements

Ben would like to thank Siân, Ivy and Poli.

Dave would like to thank Erik and Martha.

We would also both like to give particular thanks to Daryl Martin (for his supportive, brilliant and thoughtful insights) along with Steph Lawler, Laurie Hanquinet, Siân Beynon-Jones, Joanna Latimer and colleagues in the Science and Technology Studies Unit at the University of York for discussing some of the ideas that led to this book.

1

Introduction: Unpicking the Automation of Memory Making

Social media profiles inevitably leave traces of a life being lived. These biographical data trails are a tempting resource for 'platform capitalism' (Langley & Leyshon, 2017; Srnicek, 2017). As they have integrated themselves deeply into everyday routines and interactions, social media have captured a wealth of biographical information about their users. The production and maintenance of profiles has led to the recording and sharing of detailed documentary impressions. This accumulation of the day-to-day has led to the conditions in which prior content can be readily repurposed to suit the rapid circulations of social media. Moving beyond their initial remit as communication and networking platforms, social media have expanded to become memory devices. As people's lives are captured, social media platforms continue to seek out ways to recirculate these traces and to render them meaningful for the individual user. The archive is vast, and so automated approaches to memory making have been deployed in order to resurface this past content, selecting what should be visible and rendering it manageable. It is here that this book makes an intervention – this is a book about algorithmic memory making within social media. What is particularly important, as we will show, are the ways that social media's *automated* systems are actively *sorting the past* on behalf of the user.

In a short fragment composed around 1932, a piece that went unpublished in his lifetime, Walter Benjamin wrote of the 'excavation' of memories. Memories, the fragment suggests, are something to be actively mined from the continually piling remnants of everyday life. Memories require action, he implies; they are something to be achieved, they are the product of active labour. As a result, digging metaphors permeate Benjamin's single paragraph of text. He pictures the individual pursuing their memories as a kind of archaeologist combing through the dirt to uncover and reveal the items below. He opens by claiming that, 'Language has unmistakably made plain that memory is not an instrument for exploring the past, but rather a medium. It is the medium of that which is experienced, just as the earth is the medium in which ancient cities lie buried' (Benjamin, 1999a: 576). Memories, Benjamin proposes, mediate those things that have been experienced. Indeed, the mediating properties of memory was something that preoccupied Benjamin during the 1930s as, shortly following this fragment, he worked on his autobiographical book *Berlin Childhood Around 1900* (Benjamin, 2006) and later completed his famous theses on history at the end of the decade (Benjamin, 1999b). The way that we see the past through the present and the way we engage with memory and biography were important features of Benjamin's writings, especially as he sought to understand how these might be questioned and understood as political and dynamic processes.

Beyond the way that the present mediates the past, which in itself would take a book-length project to explore, it is the image of the *buried memory* that Benjamin paints in his 1932 fragment that is our main concern. For the purpose of our book, we want to simply focus on that one image and use it as a starting juncture for thinking about how memories are engaged with today. Developing that image further, Benjamin adds that, 'he who seeks to approach his own buried past must conduct himself like a man digging'. He tells us that repeated visits to the same site are often needed, and that it is 'useful to plan excavations methodically' while it can also be fruitful to do some 'cautious probing of the spade in the dark loam' (Benjamin, 1999a: 576). For Benjamin, individuals are highly active in finding, selecting and retrieving memories – indeed, it is this very process that defines a particular past moment as a memory. The fragment closes by suggesting that for 'authentic memories' it is important to 'mark, quite precisely, the site where he gained possession of them'. The vision is of the individual actively and under their own direction digging around to locate the memories among the detritus of experience. It is this active digging that gives the memory authenticity for that individual. But what if the job of digging for memories is done for us by automated systems? What happens when our memories are being sorted on our behalf and then targeted at us? What happens when we are not digging, arranging or marking out the memory spaces for ourselves, and instead it occurs as a result of algorithmic systems? Taking a step back, the question we will consider is how these processes of digging become algorithmic and what this will mean. We also seek to understand the very processes behind the automation of the excavation and targeting of memory.

Walter Benjamin's comments give us only a starting point. We use them here as an impression of how memory might be understood. Much has changed since he wrote that fragment. We do not wish to capture all of those changes to memory making in this book, although these varied technological shifts clearly include the emergence of 'digital memories' that accompanied the computer systems that have expanded in usage over the last half century or so (see Garde-Hansen et al, 2009). Rather, we wish to engage with a very particular and relatively recent development that has vast implications, the significance of which is likely to take on increasing significance in the coming years. We will not delve into the wide range of ways that memory might have been reshaped by technology;

instead we will focus on the automation processes that have been brought about by algorithmically defined social media platforms. As such, this book is concerned specifically with the way that social media content comes to be repackaged as a 'memory' before being targeted at the social media user. More than this, however, this book seeks to understand and conceptualize the way that these processes of repackaging content as memories operate and how they might be understood. Breaking down the broader developments in digital memories to think specifically about their targeting, this book examines the way that memories are *classified* and *ranked* within social media platforms. Understanding how these classification and ranking processes operate is a crucial and foundational step before we might fully understand the broader questions about memory and selfhood that social media now pose. As a so-called 'memory' pops up on our social media feed – telling us, or more likely showing us, what we were doing that day at some point in our past – we might wonder how it got there, what form of automation led to its selection, and what it might mean for the way we remember.

As this might suggest, this book will argue that with social media and other related platforms something quite different is unfolding to the vision conjured by Benjamin. Rather than digging for memories, these prominent media now seek to *do the digging for us*. Rather than requiring us to rummage in the dirt, these media aim to automatically select and deliver pre-packaged memories. We keep our hands clean. The machine does the work. We may still dig if we wish, but we also have platforms that use the vast accumulated details of our lives, as held in our profiles or in the galleries of our devices, to identify and make visible what they themselves often refer to as 'memories'. We should be very clear here about the labelling of these memories: it is the very act and politics of labelling content as memories that we are seeking to examine. We are not assuming that because social media package this older content as 'memories' that they automatically become *memories* (in a more traditional sense) for the targeted recipient. However, we are interested in how those particular bits of content come to be labelled as memories (which we focus on in detail in Chapters Two and Three), and how people respond and react to those packaged memories as they alter and intervene in an individual's relationship with and understanding of their own biography (which we begin to explore in Chapter Four). Where Nigel Thrift and Shaun French (2002) once wrote of 'the automatic production of space', we are talking here of *the automatic production of memory*.

The mediation of memory

In separating out this automatic production of memory, it is important to note that a good deal has already been written about the ways in which digital technologies affect and shape the conditions of memory practices in contemporary life (see for instance Hoskins, 2016, 2018; Keightley & Pickering, 2014; Neiger et al, 2011; Parikka, 2018). Notions such as 'digital memories' (Garde-Hansen et al, 2009) and 'mediated memories' (van Dijck, 2007) have been proposed to make sense of the ways that digital objects, such as images, social media posts, and digital archives, figure as vehicles for people's engagement with the past, both personally and socio-culturally. Together, these various perspectives provide the means for understanding how the mass expansion of memory space in computation can reconfigure individual and collective memory (see for instance Hui, 2017; MacDonald et al, 2015). For José van Dijck (2007: 21), mediated memories are a way to explore both 'the activities and the objects we produce and appropriate by means of media technologies, for creating and re-creating a sense of past, present, and future of ourselves in relation to others'. The possibilities of digital memories come to intervene in the way that the past is made and remade, which in turn feeds into notions of the self. Within this framework, mediated memories relate to the active processes involved in what Annette Kuhn (1995: 157) has called 'memory work', understood as 'an active practice of remembering which takes an inquiring attitude towards the past and the activity of its (re)construction through memory'. At the same time, mediated memories also foreground the larger digital infrastructures that facilitate their production. This means that mediated memories in the context of digital technologies are not simply what people produce nor that which simply aids memory; in reality, they have a more potent and active presence. As van Dijck (2009: 158) has pointed out, mediated memories can be seen as 'amalgamations'. It is, van Dijk adds, 'at the nexus of mind, technology, and perceptual and semiotic habits that mediated memories are shaped'. The intersections of technologies, data and memory also highlight ways in which social memory is fundamentally networked and connective (van Dijck, 2007, 2010) as well as globally distributed (Garde-Hansen, 2011; Reading & Notley, 2018). This notion of amalgamations emphasizes the entanglement of technologies, algorithmic systems, practices and human reflexivity in the production of mediated memories. This is something we will pick up on and seek to develop in relation to the specifics of social media memory work in later chapters.

One thing we should note is that we are not just dealing here with memories themselves, but with the notion of what a memory is and how the concept of memories might be defined. Following calls for more critical research into the exact ways ideas of memory are shaped by emerging technologies (Hoskins, 2018; van Dijck, 2010), scholars have recently also examined how smartphones (Hand, 2017; Özkul & Humphreys, 2015) and social media platforms (Bucher, 2018; Humphreys, 2018; Prey & Smit, 2019) shape the conditions of possibility for encountering, negotiating and remembering the past. There have even been calls for research into the ways in which platforms specifically 'facilitate memory work through the reminding of previous traces' (Özkul & Humphreys, 2015: 363). We are responding to such calls.

This book acknowledges the need for a more specific and sustained engagement with the conditions of contemporary memory making in the context of social media. Yet, with the increasing proliferation of memory features such as Facebook Memories, Snapchat Memories, Instagram Throwbacks and the like, mediated memories are no longer simply the activities and the objects that are produced and appropriated in routine practice. In addition, they have become data points that are mined, analyzed, classified, ranked and, ultimately, targeted at users. It is therefore crucial to investigate the processes by which past content comes to count as 'memories' on social media.

It is understandable that we are yet to fully explore the ways that mediated memories are automatically targeted directly at individuals and what might be the social implications of this. This is largely due to their relatively recent presence. This development may be recent, but it is already widely implemented in different forms and across different platforms. Through analyses that pay particular attention to the underlying software processes of the memory feature Facebook Memories, which we treat as a kind of archetypal version of the type of features that are already embedded across platforms and devices, we demonstrate the ways in which processes of ranking and classification are implicated in the automatic production and targeting of mediated memories. It is here, we suggest, that we can further understand memory today while also gaining greater insight into the pervasive power and politics of platforms (Gillespie, 2010; van Dijck & Poell, 2013). As such, we are seeking to build upon and contribute to the critical literature on digital memory, yet these approaches could not have foreseen the scale of the automation of memory that is now typical of social media platforms (see Jacobsen, 2020).

Metrics and targeting: the context

Before exploring this further, and before providing more detail on the specifics in which we are working, it is worth reflecting on the contextual factors that may be associated with this type of approach to social media memories. Essentially, the types of memory-making processes we discuss in this book can be associated with two broader sets of developments. On the one hand, there is the expansion of metrics and metric-based forms of social ordering, which has been referred to elsewhere as 'metric power' (Beer, 2016). On the other hand, there is also the mass expansion of targeting, which is where data are used to try to direct content toward individuals (for a summary and discussion of online targeting, see Beer et al, 2019). These two broader sets of developments are, of course, closely related. Still, reflecting on these allows us to contextualize the more specific issue of how memories come to be classified, ranked and targeted at users. Keeping these two broader contextual factors in mind allows what is happening with memories to be seen as being a part of a cultural shift toward a logic of measurement, prediction and automation. In such a context, even something as intimate as memories become something that can be calculated for their relative utility and measured for their particular value.

The power and presence of metrics in social ordering processes is now fairly clear. We are experiencing what Btihaj Ajana (2018) has described as a 'metric culture', defined by the various ways that we are tracked and turned into metrics (see also the interview with Ajana in Ajana & Beer, 2014). *Metric Power* (Beer, 2016) explored how the long history of the social application of metrics has seen an explosion in recent years, with social media and trackable devices extending the reach of what can be turned into numbers. In *Metric Power*, it was suggested that there are some key powers that metrics can be used to exercise, and that these in turn shape how we judge and how we are judged for value and worth. In the case we are describing in this book, as we will go on to explain, metricization and ranking are used to judge memories and to allow them to be ranked for their asserted value.

Of course, the presence of these metrics is not sudden, it has a longer history that has seen an acceleration in the last two decades. In his recent book *How We Became Our Data*, Colin Koopman (2019) examines how some of the recent forms of data and metric-based power were established around a century ago:

> In taking a longer historical perspective that builds out some of the back end of social media and other contemporary information technologies, I argue that the terms of our informational selfhood do not belong to social media, digital media, and other 'new' media alone. Rather, what has proven powerful about social media is their success in leveraging a particular predefined composition of selfhood that was itself designed long before anyone ever dreamt of the internet, let alone the latest darling web app. (Koopman, 2019: 7)

Despite this location or setting of a data agenda back before social media, this does not mean that nothing has changed with the arrival of social media. Social media's use of metrics and data builds upon a long-established logic, underpinned by predefined notions of the social world. The traces of the earlier data formats might be there, but the automated capacities of social media, we would suggest, mean that different things might be done within those data formats. Still, it is important to point out, as Koopman does, the effects social media may have on ideas of selfhood and identity, which in turn highlights the productive and generative potentialities of metrics and measures (see Beer, 2015; Espeland & Stevens, 2008).

Whatever the history Koopman outlines might have involved – the gaps in the printed forms, the categories, the tick-boxes, the tables, the new measures and so on – the current context sees metrics used widely as an ordering tool across various sectors. Take for instance Phoebe Moore's (2017) revealing accounts of the quantified worker, in which workplace practices and behaviours are increasingly monitored and turned into metrics against which people can be judged.

As workers are increasingly rendered into quantifiable points of comparison, it has been said that we live in a 'world of indicators' (Rottenberg et al, 2015). As a result, not even something as personal as memory can escape from the reach of these kinds of indicators. Crucially, it is the specific role of metrics in facilitating ranking processes to which we will give further attention in Chapters Two and Three. It is evident that social media platforms are metricized spaces in which interactions become data, and in which the ranking of different types of data frequently occurs. Indeed, it has been observed that social media is based on what has been called a 'like economy' (Gerlitz & Helmond, 2013) and that content is subject to the visibility regimes of algorithmic prioritization (Bucher, 2012b). In short, then, social media need to be seen within the context of what Andrea Mubi Brighenti (2018) has called 'the social life of measures', and therefore should also be seen as being part of the 'measure–value environments' that Brighenti describes. These environments are typified by the tensions and relations between measurement and measures on one hand and value and values on the other. Measures are not, Brighenti (2018: 25) argues, 'simply tools in our hands, they are also environments in which we live'. In the particular measure–value environments of social media, we are interested in how memories are measured and then judged for their value. We also want to examine some of the tensions that these environments may engender in everyday life (see Chapter Four).

As we have already suggested, increasing aspects of everyday life are 'abstracted' into data (Kitchin, 2014). The result of this, Koopman (2019) has observed, is that we become 'fastened' by our data, by which he means that they speed things up while also holding us in place. One particular aspect of the data-informed social world is the presence of targeting, where an individual's data are used to direct particular content directly towards them via their news feeds or by other direct means (described in Beer et al, 2019). This use of 'trace data' (Wu & Taneja, 2020) takes many forms: from recommendation systems and notifications through to tailored and personalized content, targeted advertising and the filtering of social media content to which we are likely to react. These algorithmic sorting processes enable content to find the individual, rather than the individual looking for that content. These data feedback loops facilitate culture *finding us*. The visibility of culture, in this view, is increasingly based on algorithmic systems deciding what individuals encounter on a daily basis and what forms of culture algorithmically finds its audience (see Beer, 2013: 97). Such targeting processes seek to be predictive and also to shape or guide behaviours, actions and choices (discussed in Beer, 2019). The impact of social media–based targeting for politics and citizenship is an area that has garnered particular attention (see Bartlett, 2018; Miller, 2019), especially as political content is targeted directly at individuals based upon the data gathered about them and as automated content curation shapes feeds (Andrejevic, 2020: 46–7, 68–9). As a result, such automated targeting also raises questions about how prejudice and inequality are coded into the 'algorithms of oppression' that have been associated with search engines (Noble, 2018) while also asking how it might extend and continue 'digital inequalities' (Eubanks, 2018; Robinson et al, 2015). These are vital issues for understanding the implications of targeted content of all forms, including memories.

The broader issues around targeting have gained attention, but there is still some scope for the particular aspects of targeting to be examined, especially around the targeting of different types of content on social media. It is here that this book aims to make a contribution, by addressing the question of how memories themselves are part of this targeting. Instead of culture 'finding us', we examine how targeted memories algorithmically *find us* on social media platforms. By focusing upon this widely employed practice in social media, we reveal something about the particular issue of the automatic production of memory on social media while also, we hope, adding something to wider debates on metricization and targeting.

The self in data

Of course, metricization and targeting often mix together. Considering metric-based calculation and ranking alongside the use of data-informed targeting produces a series of questions. One of these sets of questions concerns ideas of selfhood and person-making. In other words, these broader features of our densely connected and data-rich environments create questions for how identities form and how we understand ourselves and others. As we will go on to discuss in this book, a key aspect of this is the automatic interventions that are made in memories and biographies by these targeted systems that, as we will show, are actively categorizing content as memories and then ranking them for their relative worth and utility.

When it comes to the relationship between us and our data, one commonly used notion is the 'data shadow' (for one example see Rouse, 2016). The idea is that the data generated as a by-product of our everyday lives form into a kind of shadow that follows us around. We exist, so this vision goes, as an isolated entity that produces a data imprint. We leave an impression in the data structures we are surrounded by and that impression continually shadows our presence. A recent TV advert from the data company Experian encouraged the viewer to 'meet their data self' (Experian, 2018). The character depicted in the TV advert was replicated by an 'inseparable' clone of themselves. The advert encouraged the viewer to acknowledge this clone and to get to know their data self. Indeed, we are often encouraged to get to know ourselves through our data, or to get to know ourselves through these kinds of shadows, clones, or reflections of the self. But can we really separate *our selves* from *our data* in the way that this might suggest? Is it really viable to speak of a shadow or reflection of a stable self, or are these data more active in processes of self making? The extent to which our data shadows cling to us in contemporary society has led some to argue that 'there is nothing to disconnect from in the digital world' (Bucher,

2020: 1). When we talk about memory in the following chapters, we have to wonder if there is memory outside of these data–self relations.

Deborah Lupton's (2020) recent book *Data Selves* further explains why the data shadow type understanding of the relations between data and selfhood misses something crucial. Lupton's central premise is that the self does not exist outside of data. In contrast, the idea of a data shadow or a separate data-self lacks an understanding of how data and selfhood are entangled. The Experian advert suggests that we cannot be separated from our data-self, yet the question Lupton poses is whether the self is actually produced and shaped by data in the first place. It is therefore not so much a question of separation as one of the making of the self in the data context.

Lupton pushes towards a much more active or symbiotic understanding of the role of data in selfhood. For instance, as Lupton (2020: 12–13) writes:

> I argue that concepts of selfhood, identity and embodiment and how they are enacted with digital technologies as part of everyday life are central to understanding personal data experiences. As the title *Data Selves* suggests, I take an approach that views people and their data as inextricably entangled in human-data assemblages. These assemblages are configured via interactions of humans with other humans, devices and software, as well as the multitude of other things and spaces they encounter as they move through their lives.

When viewed as inextricable interconnections, the separation of self and data, like many other binaries, begins to break down. For Lupton, it is the entanglement of ourselves with our data that is the focal point, with an exploration of the experience of data being an important part of the equation. We may live in a vast data assemblage – which mixes human and non-human – but we all have separate and individualized 'personal data experiences' (Lupton, 2020: 12). The responsiveness of the data assemblage leads to the personalized aspects of these experiences that are also a crucial part of how data enter into processes of self making.

A key problem that Lupton identifies is that more immaterial or detached views of data mean that they are often 'dematerialized', viewed almost solely in abstract terms. The consequence of this is that it becomes difficult to see their material and active presence in social and individual life. Lupton (2020: 14) focuses instead upon the 'engagements' and 'relationships' between the 'human and non-organic' things that make up the data assemblage. Rather than dematerializing these relations and engagements, Lupton (2020: 14) seeks to see how they are 'infused with vitalities and vibrancies'. Lupton turns to the term 'more-than-human' to do this, which enables the 'human-data assemblages' to be 'viewed as ever-changing forms of lively materialities' (Lupton, 2020: 14). The assemblages themselves shift and change and so do the relations that constitute them. It is this approach to data, seeing it as 'more-than-human', that Lupton argues enables us to see aspects of data relations that could otherwise go unnoticed. This is an approach that sensitizes the analysis to the power, ethics and vulnerabilities in our data relations as 'people and their data make each other' (Lupton, 2020: 121).

In *Data Selves*, Lupton uses three particular concepts in advancing the core focal points for examining these intimate and integrated data relations: materializing, doing and sharing. These three concepts are useful for reflecting on the way that automated social media memories may become part of data selves. The first, 'materializing data', is focused on how apparently immaterial data have material implications and outcomes. Lupton's (2020: 44) argument here is that 'these materialisations have profound implications for how we make, think and feel about data, how we make sense of our data and how human-data assemblages intra-act and are made to matter'. The first area of focus is upon the way that data become material in people's lives and how data is variously made to matter in the social world.

The second concept, 'doing data', is more concerned with trying to understand the way people relate to their data and the active presence that those data have in shaping experiences. Doing data, Lupton (2020: 74) explains, is about the 'performances, enactments and sense-making work in which people engage as they respond to and incorporate their data into their lives, as they enact their data selves'. The result of this way of doing data within these data assemblages is, according to Lupton, a form of selfhood that is 'dynamic, responsive and distributed'. This area of focus is concerned with what people do with data and what data does to them. How data are received in everyday life can have a generative effect on how data selves are perceived and enacted.

The third and final concept is that of 'sharing data'. This concept explores how sharing feeds into the data that are extracted about people and also how data mediate social relations and move through networks. Indeed, this third concept draws the analysis towards a greater understanding of how data move through the social world and how they are defined by different types of social connections. This is about data's social dynamism and mobility.

As this might suggest, we are taking the position that data do not simply shadow individuals, they actively constitute and bring into existence. In the case we are describing here, we are not then thinking of social media memories as a part of a data shadow that follows people around, but rather that they have an active presence that shapes the aspects of biography that are encountered and recalled. This is to move away from seeing data simply as abstractions and to instead see them as deeply integrated into and enmeshed with human lives. The approach being outlined is encapsulated in Lupton's (2020: 122) claim that 'these data are not inscribed on bodies: they work with and through bodies'. If we are to think in such terms, then it becomes obvious that we cannot simply be separated from our data. These data are too routinely bound up with what we know of the world, how judgments are made, how decisions and tastes form, and how things are ordered and organized. The question that Lupton's book poses to us is the extent to which the data assemblages we live among and within are actively shaping selfhood.

As we think about these kinds of issues of 'the data self' to which Deborah Lupton refers, or what Koopman (2019) calls the 'informational person', then clearly the ways in which social media content is datafied and delivered back to us is a crucial part of these processes, especially if we think that memories form an essential part of identity. If we are reflecting on how people and data make each other, then the processes contained in the types of memory features we discuss in this book are a crucial part of this. Data-selves, to use Lupton's term, are likely to be at least in part shaped by the way that data intervene in memory making. And, as Lupton recommends, this requires reflections on the materialization of data, the doing of data and the sharing of data. As we will go on to discuss, data are *materialized* in the packaged memory, they are *done* in the reception of the memory and they circulate as those memories are *shared* (or targeted). Encountering this broader picture is therefore fraught and very wide-ranging. In the chapters that follow, we will focus primarily on the first of these steps and on what Lupton refers to as data materialization as it occurs in the form of social media memories. This takes us to the devices themselves, particularly Facebook's throwback feature, which is simply called Memories.

Engineering memories through throwback features: the case of Facebook Memories

Most social media users are likely to have encountered a social media 'memory'. Even infrequent social media users may have encountered such content. These repackaged bits of past content are presented in highly casual tones and can even find their way into our interactions in indirect forms ('Look at this picture of us on holiday three years ago today') as well as more direct appearances in social media news feeds ('You have a memory with Steve to look back on today'). However, as we have indicated, our concern is largely with the processes that lead to the delivery of such content. In other words, how does that content come to arrive on the screen. It seems necessary, for us at least, to clarify what type of input and formalization of content is needed for this automatic memory making to function, as well as analyzing the various possible responses that these packaged memories might stimulate. As the book seeks to show, the sorting processes are the foundation from which it is then possible to begin to understand how memory might be reshaped through automation. However, we are not thinking solely in technical terms. We are also approaching these systems as socio-cultural formations. Automated social media memories can be understood to be a part of the current 'memory ecologies' (Hoskins, 2016), where technologies, platforms, environments, objects, individuals and groups are increasingly entangled and where both personal and collective memory is facilitated. Arriving at the type of automatic memory delivery that is seen on social media is a product of a range of relations that mix with certain ideals and notions of biography, selfhood and remembering.

Let us be more specific. What are sometimes referred to as throwback features are essentially mechanisms for presenting previous content to a social media user, usually at the point of a particular anniversary of the date when a given 'memory' was posted or uploaded. From June 2018, Facebook's throwback feature was called Memories. We take it as a kind of archetypal example of these varied social media throwback features. Although it can be considered one of Facebook's many supplementary features, Memories forms an integral part of many users' experience of the social media platform, with an estimated 90 million visits each day (Reid, 2019). Indeed, with the emergence of similarly popular tools such as Apple Memories and Snapchat Memories, throwback-type features are becoming much more prominent in social media and devices more generally. Facebook's Memories feature, along with 'Groups' and 'Pages', can be found on the left-hand side of Facebook's News Feed, and it consists of past content such as posts and images that are updated according to what has been posted and when. It enables users to revisit content from a given day in their Facebook history. Memories may also appear in a user's News Feed regardless of whether or not they intentionally visit the feature. If no memories are shown on a given day, it means that either a user did not post anything on that particular day in the past or that, as Facebook put it, 'We don't have anything to show for that particular day' (Facebook Help Centre, 2018). What then becomes visible in the feature, that is to say what memories emerge, is depicted as a collaboration between a user's sustained participation on the platform and the specific patterns the feature is able to extract from that data.

When launching the feature, Oren Hod, Product Manager at Facebook, wrote: 'Today, we're launching Memories, a single place on Facebook to reflect on the moments you've shared with family and friends, including posts and photos, friends you've made, and major life events' (Facebook Newsroom, 2018). Hod states that the rationale behind the feature is based on research from people using On This Day, an earlier iteration of Memories. 'Research suggests', Hod continues, 'this kind of reflection can have a positive impact on people's mood and overall well-being. This is why we're updating the experience to ensure all of your memories are easy to find' (Facebook Newsroom, 2018). Clearly this is being presented in positive terms of wellbeing and personalized experience, as a form of interaction with past content that can be integrated into general social media use. This was also echoed by Artie Konrad, a researcher involved in the development of the feature. In an interview with the American Psychological Association, titled 'Jogging happy memories', Konrad stated that 'our goal is to create a delightful experience' (Reid, 2019). Indeed, Konrad's earlier work was in the field of psychology, researching the potential long-term impact of technology on memory and well-being (see for instance, Konrad et al, 2016). In the interview, Konrad stated that 'our research showed that reminiscing with technology can reconnect people with positive emotions and increase long-term well-being' (Reid, 2019). The suggestion here is that there is a push toward delivering memories that are considered positive, uplifting or that add to the users' overall 'well-being'.[1]

In all of this, we see hints of a particular definition of memory that pushes toward particular types of memory selection. Such an approach would have a prerequisite and ability to categorize those memories that fit into that model. On some levels this assertion seems obvious; we might ask who would like to use features that have a bad or negative impact on them. But, of course, existing definitions and approaches to memory capture something much more varied in terms of sentiment (see for instance Kuhn, 1995; Lawler, 2008; van Dijck, 2007). Further to this, we can see a specific logic of positivity and wellness coded into the feature's core aims and rationale. According to TechCrunch, and echoing Hod and Konrad, the Memories feature is intimately tied to Facebook's 'time well-spent' efforts that aim 'to increase the focus on quality engagement on Facebook' (Constantine, 2018). Facebook's new throwback iteration is therefore conceptualized within a very specific framework, one that establishes relevancy in terms of 'time well-spent' along with an assessment of what is considered conducive to sharing and increased user engagement on the platform. We begin to see here the notions of value upon which memories might be classified and then ranked. The ideas that drive these innovations arrive with some specific parameters and ways of thinking. As we will show in Chapters Two and Three, this underlying logic also participates in demarcating people's field of visibility on the platform in particular ways.

If we turn for a moment to the type of content that Facebook's Memories function seeks to deliver, we can identify further attempts to circumscribe the types of memories that are seen as valuable or desirable. What memories, then, can users expect to see resurfaced with the feature? According to Facebook Help Centre (2018), one can expect to see the following things on Memories:

- Your Facebook posts
- Posts on Facebook that you were tagged in
- Major life events
- Your marriage anniversary (if you've added it to your profile)
- When you became Friends on Facebook with someone
- When you joined Facebook
- Photos from your mobile device's camera roll
- Recaps from the past month or season.

The feature includes both sections that may seem mundane or playful (such as 'When you became Facebook Friends' or 'When you joined Facebook'). Other sections comprise more intricate arrangements of data. The section 'Recaps of Memories', according to Oren Hod, will feature 'seasonal or monthly recaps of memories that have been bundled into a message or short video' (Facebook Newsroom, 2018). Likewise, the section titled 'Memories You May Have Missed' includes posts from the past week a user might have missed in case 'you haven't checked your memories lately' (Facebook Newsroom, 2018). By including multiple sections within Memories, the overall aim is to increase the likelihood that users will have 'quality engagement' with some of these things and the likelihood that participation on the platform increases. Including more sections is also aimed at producing an even more personalized experience of the platform. This pursuit of personalization of experience is reflected in Oren Hod's claim that 'we work hard to ensure that we treat the content as part of each individual's personal experience' (Facebook Newsroom,

2018). The delivery of memory is now a prominent part of the attempt to extend personalization still further (Prey & Smit, 2019).

This general overview of the functionality of the Memories feature indicates that the platform deploys a set of predetermined criteria for what constitutes a 'memory', criteria that are partly based on the research previously conducted on people using Facebook's older iteration, the On This Day feature (we will discuss the circumscription of memory in more detail in Chapter Two). As such, Facebook's conceptualization of what a memory is is underpinned by a particular logic. It is a logic that frames reminiscence as personalized, positive, and as holding and foreshadowing a particular type of future engagement. This particular logic, however, only extends to certain parts of a user's past on the platform. We can therefore ask what the logic of wellness and positivity obfuscates. For instance, Facebook attempts to automatically detect and identify memories that people prefer not to see again. Based on factors such as reactions (emojis for instance) and keywords ('Sorry about the break up') these memories are filtered out and thus not resurfaced in the future (Carman, 2018; Paluri & Aziz, 2016). As a result, we see that a kind of sentiment analysis permeates the definition of what is considered to be a memory, or at least what is considered to be a valuable memory within the logic of the platform. The remits of the feature, of what constitutes a 'memory', are demarcated relatively clearly, and according to a specific logic. In keeping with the broader 'happiness industry' (Davies, 2015), in which sentiments and emotions are carefully managed and shaped, the field of the visible past on Facebook Memories is conceptualized through the prism of what adds to a user's overall 'wellbeing', which in turn excludes and renders irrelevant other parts of a user's past. This represents a purification of selection toward automatically jogging only narrowly conceived notions of happy or positive memories. There is only room, it would seem, for happy memories.

A note on the direction of this book

The automatic production of memory that we are dealing with represents a broad set of changes with wide-range of implications. This type of automated personalization, including the targeting of memories, was not fully possible prior to the presence of social media and the traces of everyday life that accumulate within them. In the following chapters, we will look at how the automated selection and targeting of memories is based upon a dual process in which social media features actively and autonomously classify *and* rank people's past content to produce and deliver these ready-packaged 'memories'. What we find will be more of these embedded notions of value and worth, along with attempts to define and redefine the very idea of memory and memories. Where Benjamin pictured a digger or archaeologist of memory, we use this book to explore how memories are classified and ranked so that they can be unearthed on our behalf. Neither planned nor aimless excavations are needed where the memories are already selected and polished within our social media feeds. In short, the digging has become algorithmic.

Our view here is that this dual process of classification and ranking has profound consequences for what memory is and how it forms, as well as having implications for notions of collective and individual memory on a larger scale. Flavoured by the broader objectives and agenda of social media platforms, the way that memories are automatically produced through the processes of classification and ranking will inevitably have an impact on the memoryscapes that we are presented with, shaping what is remembered and how. There is a need to ask what happens to memory when there is no longer this need to dig. In this book we focus on how memories are what we call 'partitioned' and 'promoted' by algorithmic systems and, drawing upon interviews and focus groups, how individuals respond to those combined processes of classification and ranking.

Following this introduction chapter, the book is split into three core chapters. Chapter Two deals with the classification processes in which memories are defined and, as we will describe, partitioned. Chapter Three builds upon this by exploring how these classified memories are then ranked and promoted. That is to say it looks at how memories are brought to the surface among all the vast archives of past content. With this dual process of partitioning and promotion established, we then look in Chapter Four at the tensions of remembering that this creates. We specifically look at the tensions that emerge as a result of the classification and ranking processes behind this automated memory work. Chapter Four begins to outline the various responses and reactions people have to such processes and to the automated defining and delivery of personal memories in everyday life. As a result, this book is an examination of *the automatic production of memories* that occurs as social media extract, categorize and then target memories at us. Let us begin in the following chapter with the taxonomy that resides behind this type of memory making.

2

A Taxonomy of Memory Themes: Partitioning the Memorable

Inevitably, classification processes are powerful within any type of archive. The way content is classified shapes how documents are interpreted and, crucially, how they are retrieved. If we approach social media as a form of archive, then we can begin to see how the ordering process of classification and sorting that occur within these media may be powerful for how people engage with their past content and how individual biographies are made accessible. As we will explore, the ordering of the archive is crucial for understanding its functioning and what can be pulled from its vast stores.

The types of archives that are used to document life are powerful in their presence and outcomes. For some it has been placed at the centre of modern power formations. Derrida (1996: 4 n1) famously argued that, 'there is no political power without control of the archive, if not of memory. Effective democratization can always be measured by this essential criterion: the participation in and the access to the archive, its constitution, and its interpretation.'

If we treat social media as a population of people effectively participating within a large archival structure, then social media bring the politics of the archive to the centre of everyday life and social interaction (see Beer, 2013). Derrida's point is that the structures of the archive afford its uses and what can then be said with it or retrieved from it. He argues that 'the technical structure of the *archiving* archive also determines the structure of the *archivable* content even in its very coming into existence and in its relationship to the future' (Derrida, 1996: 17; original emphases). The form that the archive takes also dictates the type of items or documents that come to be stored within them; it imposes its logic upon its content. Derrida adds to this, crucially, that 'the archivization produces as much as it records the event' (Derrida, 1996: 17). The technical structures of the archive need to be understood in order for its politics to be revealed, particularly as they intervene in the relations between the past and the future. This is something that we will keep in focus as we move through this and the following chapter. We will use these chapters to indicate how, in Derrida's terms, social media archives and the throwback processes within them are actively producing as much as recording events. Whatever position we may take here, we would suggest that there is a clear need to engage with the archiving and classificatory processes that are occurring in social media, especially in the automated retrieval of content in the form of memories.

Within these archival structures Derrida argues that it is also necessary to understand the figures who actively create and deploy classificatory systems, deciding what is placed into the archive and what can be retrieved. Derrida talks of these figures as 'archons', claiming that:

> The archons are first of all the documents' guardians. They do not only ensure the physical security of what is deposited and of the substrate. They are also accorded the hermeneutic right and competence. They have the power to interpret the archives. Entrusted to such archons, these documents in effect speak the law: they recall the law and call upon or impose the law. (Derrida, 1996: 2)

In this passage, Derrida talks in terms of the documents being used to recall, speak and impose the law. If we approach social media as archives, then the classification systems within them need scrutiny. This is especially the case where classificatory and archival processes might impinge upon memory and upon how the past comes to matter in the present. The figures who classify and then sort content, or, as we will describe, those who partition and then surface these memories, are similarly in need of scrutiny. As we will describe in this chapter and in Chapter Three, there is scope to see so-called 'algorithmic archons' (Beer, 2020) at work in these social media archives, with automated systems actively classifying the content and conducting other archival practices. Yet, as we will also show in this chapter, these are not entirely automated processes. Rather, these algorithmic selections operate through an existing taxonomy and classificatory system that underpin the way in which content is rendered retrievable.

The power of classification in memory making

Before examining how exactly the Facebook Memories feature classifies past content, it is worth acknowledging the importance of classification for what Geoffrey C. Bowker (2008) calls 'memory practices'. In his book *Memory Practices in the Sciences*, Bowker examines the range of practices, technical and social, by which the past is committed to record. With the increasing proliferation of novel technologies, Bowker argues that memory practices remain central to our ways of knowing and understanding ourselves and others. In a word, 'they skew our available ontological space' (2008: 71). For Bowker, memory practices constitute different ways to frame the present as well as the past, which in turn shapes how the social world is perceived, understood and enacted. Bowker suggests that classification systems are not only an important aspect of how memory operates and how the present is framed.

Indeed, he argues that the idea of the past is inseparable from classificatory processes. As Bowker states, the past is 'less a record than a sort of taxonomy', adding that 'our testimony depends much less on our memory, than on the mental image that we possess of a type or a class in which we arrange facts' (2008: 16). Classification helps us understand the past in the present through its segmentation into intelligible and meaningful boxes. Indeed, the inseparability of classification and taxonomy from ordinary life has led Bowker and Star (2000: 1) to claim that 'to classify is human'.

Echoing Derrida's understanding of the archive, Bowker (2008: 30) also argues that the current epoch is one marked by the proliferation of 'potential memory' (see also Bowker, 1997). In other words, it is characterized by the increasing documentation and archivization of the past and the present. This may initially seem fairly obvious in an age of platforms and software features: we create countless digital traces on an everyday basis, most often unconsciously. Yet, the notion of potential memory, akin to Derrida's archive fever, signals the pervasive drive to capture and order the present not for any specific purpose, but 'should the need ever arise' (Bowker, 2008: 30). The pervasive documentation of the present, therefore, is not in and of itself sufficient. Traces of the present must be classified, standardized, and ultimately ordered for these to be rendered potentially meaningful in the future. In this view, to be remembered is to be classified. To be understood is to be segmented. It is to be fitted into a certain classificatory schema that renders the world knowable and meaningful as well as potentially usable. Indeed, Bowker (2008: 24–5) argues that memory must be understood as fundamentally interwoven with processes of classification:

> The act of remembering sometimes coalesces into more or less complete classifications, more or less rigorous standards. But more generally, it is one of our chief ways of being in the world as effective creatures: it is a way

of framing the present; a mode of acting. We exude, sketch, form, design memory traces of all kinds; these work together in complex ecologies.

In this view, memory practices constitute a drive to render the world memorable, which in turn emphasize the role of classification systems. As Bowker observes (2008: 30): 'In order to be fully countable and thus remembered by the state, a person needs first to fit into well-defined classification systems.' Yet, in an age of algorithms and social media, what does it mean to render the world memorable? Indeed, how are memories classified and produced in this context?

Pigeon-holing the past

Having briefly introduced the logic underlying of social media's memory features in Chapter One, we ask in this chapter what components can be said to be key in determining the classification of a 'memory' on Facebook Memories. As we have mentioned, we take this particular feature as an archetypal instance of various memory features that continue to be established across social media platforms (and are being embedded into smartphones as well). The question here is what properties are used to turn content into memories and what systems are used to order and classify the content that is then judged to represent a potential memory. This question opens up a space for thinking critically about the ways that processes of classification on social media produce certain visions of ourselves, of others, of the past, present, and future – in other words, how they shape 'our available ontological space' (Bowker, 2008: 71). Before memories can be targeted, the first step is for them to be classified and taxonomized (which then facilitates the ranking processes that we will go on to discuss further in Chapter Three). In what follows, we will outline these initial stages of the separation of content, conceptualizing these separation techniques as a *partitioning of the memorable*.

How exactly does Facebook Memories, then, determine what past content becomes a 'memory' and what does not? In an interview with Shauna Reid for the American Psychological Association, Artie Konrad, a user experience researcher at Facebook, articulated the early challenges facing the developers of Facebook when it came to resurfacing memories. Konrad stated that:

> From past research, we knew our users wanted a way to easily revisit these memories from their timelines. But when 'On This Day' launched in 2015, we didn't yet know what people wanted from the experience. What types of memories did they want to see? How often? I've since conducted 15 studies to help answer these questions. (Reid, 2019)

Konrad depicts a mutating technology that has been shaped through the data feedback loops that have been gathered and analyzed. In 2017 Konrad published a report titled 'Facebook memories: The research behind the products that connect you with your past', which aimed to reveal how Facebook had sought to practically answer these questions. The report outlined in detail how Facebook had developed their original throwback feature, which they then called On This Day, an earlier iteration of the current Memories feature. In the report, Konrad explains that before developing the feature participants were invited for interviews at Facebook's research labs, located at their headquarters in Menlo Park, California, with the aim to determine Facebook's role with memory (note here that Facebook had a predetermined notion that they had a role to play in memory, which would suggest that they were already seeing themselves as a type of archon). From this study, it was concluded that Facebook should 'provide reminders of fun, interesting, and important life moments' (Konrad, 2017; see the discussion in Chapter One). The initial issue was therefore how to define and operationalize these within the feature. As

Konrad (2017) states, 'we set out to figure out what makes a memory fun, interesting, and important'. At this point, the path was also set for the type of memory that would be valued and that the systems would seek to make most visible.

Ideals of engagement and the sorting of past content

In an example that illustrates a version of the 'social life of methods' discussed by Mike Savage (2013), Konrad outlines a three-pronged approach that was taken to determine Facebook's role in memory making: memory themes, quantitative surveys and linguistic analysis. Firstly, Konrad describes how he had the participants classify their memories into various themes such as 'vacation', 'food' and 'family', which helped generate both the scope and specificity of people's memories on the platform, and upon which later research and algorithmic models could be based. In a sense, participants helped generate distinct categories of memories, but they were also invited to formalize their memories into categories already pre-set by the researchers. The grid began to form. This, in turn, provided the social media platform with what can be considered as 'sufficient approximations' (Gillespie, 2014) of what people generally find memorable or the things they see as being most worthy of the category of 'memory'.

Secondly, the 'memory themes' that emerged from the qualitative research findings were then verified in larger quantitative surveys sent to random users on Facebook. In the survey, people were encouraged to classify their memories into the themes developed in the qualitative interviews as well as 'rate how much they enjoyed seeing that memory' (Konrad, 2017). The coupling of qualitative interviewing with a quantitative survey method provided an added layer to how memories come to be understood, defined and produced by the platform. As a result of this large-scale quantitative survey, Konrad stated that: 'I learned, for instance, that pictures of family are memory gold, especially kids because there is surprised delight in seeing how much children have grown over the years.' Clearly the tone of the memories they were seeking to 'make' was being established through the methods being used along with an established sense of what the platform wanted the user to get out of the memory.

Lastly, the researchers conducted a linguistic analysis of words within anonymized memories across Facebook in order to better understand the kinds of semantic and discursive attributes that are attached to what are considered to be 'good memories'. The linguistic analysis, Konrad states, provided insights into the kinds of past content or memories people are most likely to share on the platform. Through this approach, Konrad found that 'memories that had words like "miss" in them (i.e. "miss your face") were most likely to be shared, whereas food–related words (i.e. "best taco ever") were less likely to be shared because they were no longer relevant'. This is illustrative of how notions of relevance and likely shareability were factored into the assessment of value within these burgeoning systems. Good memories are, in this logic, shareable memories.

It is notable that at each step in the development of Facebook Memories important assumptions about memories were being made, assumptions that would inevitably then impact on the future retrieval of memories. In the qualitative interviews, for instance, participants were encouraged to frame and classify memories thematically, suggesting that memories are assumed to be classifiable in this way. At the quantitative stage of the analysis, memories were also assumed to be rankable. This helped the researchers and developers at Facebook identify patterns of relevance in the data. Again, relevance was a key means by which the value of memories was being measured and weighted. Ranking memories was a means of formulating remembering metrically, using numbers to assess the significance, shareability and relevance of people's engagements with the past. Thus, 'the desire for numbers' (Kennedy, 2016) encapsulates the ways users' memories are conceptualized, classified and ranked within the feature. At this stage, it might be tempting to consider these assumptions solely in terms of bias. But rather than constituting a hindrance, these assumptions are crucial to the feature's functionality and to the way that memories are now circulating. As Louise Amoore (2020) observes, all algorithmic systems require bias and underlying assumptions in order to function in the world. In fact, 'bias and error are intrinsic to the calculative arrangements – and therefore also to the ethicopolitics of algorithms' (Amoore, 2020: 74). The assumptions underlying the development of Facebook Memories tell us something interesting about the way in which processes of classification and ranking come to be entangled with people's understanding of memories.

After having conducted this three-pronged research approach, Konrad (2017) states that the themes and findings were aggregated and consolidated into a unified framework they called the 'Taxonomy of Memory Themes'. The taxonomy divides memories into categories that include relationships, sport, holidays, qualifications, meals, family, birthdays, pets and the like. The framework functioned as a mechanism to establish the relevance of memories on Facebook and the likelihood of them being shared. The taxonomization of people's content into different *types* of memories helped render these knowable and actionable to the feature's ranking algorithm. As Konrad (2017) observes, the framework 'not only helped us bucket memories into distinct categories, but began to inform our ranking algorithms that select which memories to surface'. The classifications then separated out the memories into groups but also enabled a more direct form of sorting in terms of the prioritization of memories within these categories.

The taxonomization therefore determined which memories are rendered archivable and retrievable and in what ways. Moreover, this classificatory framework functions as a way to 'commensurate' (Espeland & Stevens, 2008) people's past content, establishing a common standard or taxonomy by which memories could be quantified, distinguished, compared and ranked. The memories are also rendered retrievable by these sorting processes, enabling their automatic production and targeting. Processes of classification and ranking can therefore be seen to be intimately related on Facebook. It is also important to note that the ranking algorithm is based on the crucial assumption that not all memories are equally weighted, that they will not have equal (emotional) resonance with users. Seeking out those resonances was their aim. Some memories will appear to be more meaningful, in their terms, than others and thus feature higher in the ranking algorithm. In place of the piles of potential memories that need to be dug through, the classification and ranking of memories facilitate and reinforce particular forms of meaningfulness of past events and prior social media content. As Bowker (2008) once observed, the past emerges in and through classificatory systems.

Partitioning and surfacing

The main factors determining whether or not a memory will resurface are then two-fold: firstly, the resurfacing of a memory is determined by its position within the pre-existing classificatory framework and, secondly, its weight is then determined by the ranking algorithm. Through the development of the 'Taxonomy of Memory Themes', we see a step-by-step example of what Adrian Mackenzie (2015) called 'the production of prediction'. The production of prediction that we are outlining here is a process by which algorithmic predictions are made actionable within the throwback feature, and are thus trying to predict which particular memories a social media user may wish to see. Noting the specific importance of classification, Mackenzie (2015: 433) argues that 'prediction depends on classification, and classification itself presumes the existence of classes, and attributes that define membership of classes'. We are finding exactly this. The classification and ranking of memories work together to influence what is made visible and what is then consumed as a 'memory'. In short, classification and ranking are used together to produce the predictions of which memory the user is likely to wish to have resurfaced on their behalf (there is also the initial stage of predicting which bit of content the user may consider to be a memory rather than just routine content). The weight of a memory and its relevance therefore comes to be conceptualized within a pre-existing set of classifications and a calculative regime. Chapter Three will outline in more detail how Facebook's ranking algorithm measures and ranks memories.

If we leave the ranking process to one side for the moment, one of the questions raised by the processes discussed concerns the separation of memories into categories and their distribution across classifications. Of course, much has already been written about the role of classifications in shaping the way the world is perceived, understood and structured (see for example Bechmann & Bowker, 2019; Bowker & Star, 2000; Foucault, 2002). Features like Facebook Memories constitute a nascent development, where the automatic production and resurfacing of memories is predicated on how memories are defined and classified. This automated memory making adds a new dimension to existing debates on the power of classification on one hand and memory making on the other. In the case of Facebook Memories, people's parameters of visibility concerning past content are based on the sociotechnical taxonomization of their content into clearly defined categories with clearly defined attributes, which in turn renders them knowable and actionable to the ranking algorithm. Yet, what are the implications of this taxonomization for the way memories are understood by the platform? What are its implications for the conditions of possibility of people's memory making in everyday life? In order to better understand the complex role of classification in the automatic production of memory, we seek to draw on Jacques Rancière's notion of the 'distribution' and 'partition' of the 'sensible'.

Distributing and partitioning the sensible

In his famous essay 'Ten theses on politics', Jacques Rancière (2010) explores the way that politics frames and shapes conceptions of the world. The essay deals with issues substantially different from those we are covering here, yet there are conceptual moments in that piece that provide possibilities for exploring the automatic production of mediated memories on social media. For Rancière, politics constitutes processes of 'distributing the sensible' (2010: 44). This 'distributing of the sensible' is concerned with the spreading of certain ways that the social world might be seen or understood. It is here, he argues, that the parameters of perception are defined, which in turn shapes how people act in the world. Rancière (2010: 44) writes that:

> I call 'distribution of the sensible' a generally implicit law that defines the forms of partaking by first defining the modes of perception in which they are inscribed. The partition of the sensible is the dividing-up of the world (de monde) and of people (du monde), the nemein upon which the nomoi of the community are founded. This partition should be understood in the double sense of the word: on the one hand, as that which separates and excludes; on the other, as that which allows participation.

The process of distributing the sensible is, for Rancière, a field of constant contention, one where individuals and groups engage in a perpetual dissensus, that is, a struggle for the right to define and legitimate a certain conception of the social world. The distribution of the sensible and its relationship to what Rancière calls the 'partition of the sensible' play a crucial role in defining the limits of participation.

The partition of the sensible specifically refers to, according to Rancière, 'the dividing-up of the world', the establishment of the boundaries that frame participation and produce spaces of both inclusion and exclusion. As Kwek and Seyfert (2017: 37) observe: 'Partitions make us *sensitive* to particular configurations of experience; partitions selectively make things make *sense*.' We might see here how the classification of memory in terms of this distribution or partitioning of the sensible helps define what is remembered and how participation in social media then unfolds. In other words, we need to ask what configurations of memory are partitions making us sensitive to. Rancière (2010: 44) continues by arguing:

> A partition of the sensible refers to the manner in which a relation between a shared common (un commun partage) and the distribution of exclusive parts is determined in sensory experience. This latter form of distribution, which, by its sensory self-evidence, anticipates the distribution of part and shares (parties), itself presupposes a distribution of what is visible and what not, of what can be heard and what cannot.

Here the sensory experiences of the individual are shaped by processes of partitioning that foreshadow them. Rancière closes here by pointing out that partitioning processes dictate what is encountered and what is not, what is visible and audible from what is not. Partitions make things make sense. In the context of this article, we extend this notion to what is remembered and what is not. As Kwek and Seyfert (2017) indicate, we need to ask what particular configurations of memory these partitions are making us sensitive to. In other words, we need to ask how content is divided up so as to engender particular sensory experiences of the past.

Although this is a field in flux, marked by struggle and political dissensus, Rancière argued that the sensible also crystallizes. In order to better explain the political role of partitions in social life, Rancière (2010) uses the example of the police. One of the main responsibilities taken on by the police is to ensure social order in the face of unexpected or potential chaos. Rancière suggests that the way regulation and order were ensured, especially in situations where there has been a crime or an accident, was through partitioning the sensible. As the example goes, the police will set up barriers, road blocks or barricade tape while shouting to curious onlookers and passersby 'there's nothing to see here!' (Rancière, 2010: 45). For Rancière, this was a means for the police to not only ensure order, but also a way to partition the sensible, to divide between that which people should not concern themselves with and the rest of everyday life. Classifying space as respectively accessible and non-accessible, partitioning the sensible was a means to separate the chaotic from order, to control and manage uncertainty, and to loosen the accidental from the everyday, practical and meaningful. In short, partitioning the sensible is to skew people's 'grid of intelligibility' as Michel Foucault (2008: 252) once put it. Of course, the police, for Rancière, is not merely an analogy, but also a concept signifying 'the symbolic constitution of the social' (Rancière, 2010: 36). The police is that which asserts that 'the space for circulation is nothing but the space of circulation' (Rancière, 2010: 37). It is that which partitions and divides up the world. The partitioning of the sensible is therefore a highly political act, if politics is understood as that which produces 'the set of horizons and modalities of what is visible and audible as well as what can be said, thought, made, or done' (Rancière, 2004: 85). The question, then, is how social media's automatic production of memory might be partitioning the sensible, enforcing a circumscribed kind of order (as opposed to chaos) and demarcating that which should have attention from that which should not.

What Rancière points toward is the way that partitioning affords and delineates spaces of visibility and invisibility. One of the ramifications of partitioning the sensible is that it ultimately shapes what is conceived and the perceptions that spring from them, which in turn shapes how one is able to participate in the world (or participate in social media, perhaps). As Rancière (2010: 44) puts it: 'This partition should be understood in the double sense of the word: on the one hand, as that which separates and excludes; on the other, as that which allows participation.' So the partitioning of the sensible is a process that goes beyond visibility and invisibility; it is about exclusion and participation as well. For Rancière, it is a mechanism through which the sensible as a whole is variously partitioned and repartitioned, divided and classified, and ultimately rendered manageable and controllable. The process of partitioning is where decisions are made about what is rendered available and therefore what is participated within.

Conclusion: the partitioning of the memorable

Returning to the 'Taxonomy of Memory Themes' generated by Facebook Memories, Rancière's notion of partitioning of the sensible is a helpful conceptual tool to better understand the implications of this particular classificatory system. Indeed, the massed recording of everyday life on social media platforms is ripe for partitioning to bring order and render controllable the depths of stuff. Thinking through the lens of Rancière brings to the fore two critical attributes of practices of classification: the way people's grids of intelligibility are shaped through processes of classification and, secondly, the way things and experiences are ontologically skewed and reconfigured in those processes. For Rancière, when any given area is partitioned off, the frontiers of the sensible are re-formed and, crucially, the fundamental relationship between the sensible and non-sensible is reconfigured. In the context of Facebook Memories, the partitioning-off of certain memories and the availability of others fits with these broader ideas. The implications of rethinking the relationship between classification and memory making on social media through the lens of Rancière is to sensitize us to the way things – such as previous content and these labelled 'memories' – are reconfigured and come to matter in a new way on these platforms. We argue that when past data are classified and taxonomized, Facebook simultaneously establishes the conditions to determine what is memorable. Moreover, the partitioning of past content establishes the boundaries of what can and should be remembered and what should not – that is, what is deemed memorable.

Through the lens of Rancière, we see that Facebook Memories participates in the partition of the sensible in a narrower form. More specifically, we argue that social media are directly participating in *the partition of the memorable*. Instead of saying, as we saw in Rancière (2010: 45), 'Move it along! There's nothing to see here!', Facebook proclaims 'here's something to see: a memory!' At the same time, the platform is drawing attention away from the other content, which is deemed unnecessary and less valuable: 'There's nothing to see here!' (Rancière, 2010: 45). As such, the platform shapes people's grid of intelligibility through the reconfiguration of past data into 'memories'. It partitions what is deemed memorable as well as specifying the type of memory through processes of classification and, as we will discuss further in a moment, ranking. The classification of past content as 'memories' is an attempt to render them meaningful and affective in the present, while also excluding content not deemed memorable by the platform. As a result, the partitioning of the memorable has the potential to shape the conditions by which content is resurfaced and encountered in everyday life, shaping what *can* and *should* be remembered, celebrated and shared. Partitions seek to make the past make sense in selective ways.

What we have described in this chapter is how past content comes to be labelled as a 'memory' within the logic of the social media platform. Through processes of classification and taxonomization, Facebook partitions the memorable, determining what content is deemed memorable and therefore resurfaceable and what is not. The automated partitioning of the memorable in social media, through processes of classification, sets up a new regime of remembering and forgetting. This returns us to Derrida's politics of the archive, which we can see being exercised directly, by algorithmic archons, in attempts to intervene in the memory practices and memory making of the individual social media user. The way these classificatory systems operate is crucial for understanding the way that memory is conceptualized, sorted, promoted and targeted within social media. As this chapter has shown, memories come to matter in the present through classification. Through such throwback or memory features, social media are actively drawing upon the archives of individual users while acting as archons in their automated decisions about which category a piece of content might fit into. We then need to ask how memories, classified and partitioned, come to be resurfaced and targeted at users. It is to that question of surfacing and what we call 'promotion' that we will now turn to in the following chapter. If this chapter has sought to show how memories come to matter through classification and partitioning, the next chapter seeks to show how they are made to matter through processes of ranking and promotion.

3

The Computational Surfacing of Memories: Promoting the Memorable

Although we are talking about the automated production of memory in this book, these systems are still anchored by classification systems that open them up to a much longer held and well-established, as Foucault (2002) put it, *order of things*. It is also important to note that 'The Taxonomy of Memory Themes' discussed in Chapter Two served as the 'ground truth' (Amoore, 2020), so to speak, for the development of Facebook Memories. Established prior to its development, the memory classifications generated by Facebook's research studies were fed into the design of Facebook's current throwback feature. This was effectively a moment in which the formalization of a computational problem occurred and where there was an attempt to render the indeterminable and contingent into something calculable (see Fazi, 2018). Once this taxonomy of memories was in place, it provided the ranking algorithms with a clear-cut computational problem to 'solve' and optimize: *what* to surface, to *whom* and *when*. In other words, once there was a system in place for classifying memories within the taxonomy, the system had to then decide which memory, from all these many classified memories, should be targeted at the intended recipient and when they should receive it. Once the classificatory system is active within this social media archive, the focus then has to shift to retrieval and to the way in which this retrieval is instantiated in processes of ranking. Bringing memories to the surface requires, in this logic, a system by which they can be ranked – memories ranked at a certain level are the ones that then become visible. It is this ranking of memory that this chapter deals with.

Feedback loops and the surfacing of memories

In a Facebook Research report titled 'Engineering for nostalgia: building a personalized "On This Day" experience', Manohar Paluri and Omid Aziz (2016) outline the software engineering side to building the earlier iteration of Facebook Memories called On This Day. The claim behind this, they explain, is that they 'wanted to make sure On This Day shows people the memories they most likely want to see and share, especially when it comes to the memories they see in News Feed' (Paluri & Aziz, 2016). As with other algorithmic systems, there is an attempt to predict the optimal content individual users will want to consume at any given point. In this case, the optimal content to be predicted constitutes users' own past content repackaged as 'memories'. Paluri and Aziz mention that in order to optimize people's experience of the feature, they focused on three main areas of development: user experience research, filtering and ranking. As Chapter Two showed, Facebook's user experience research generated a framework called the 'Taxonomy of memory themes' through which memories could be classified and grouped (Konrad, 2017). Secondly, Paluri and Aziz (2016) sought to optimize the memory feature by developing automatic filters. These were implemented to try to automatically filter out memories with negative associations attached to them: images depicting painful experiences, relationships that had since broken down, accidents, people who might have died and so on. As we pointed out in the introduction, such memories run counter to the platform's underlying logic of wellness, positivity and 'time well-spent'. As we also showed, this filtering process was actualized through the identification of metadata such as image captions or comment sections and the use of emojis.

Facebook's taxonomy of memory themes worked to not only establish classes of memories, but also to clearly delineate the attributes and discrete characteristics of particular classes of memories. Conversely, here we see the filtering out of certain memories based on the identification of attributes considered to have a negative impact. Paluri and Aziz (2016) add that this filtering was also accompanied by attending to users' own preferences, that is, the specific people, images or dates that users have themselves filtered out. The optimization of filtering memories was therefore partly derived from mining and understanding the attributes of users' own filtering practices, attributes that were in turn fed to Facebook's algorithm. Here we see the other side of the visibility produced through the ranking of memory, with the ranking algorithms actively producing invisibility for certain memories (see also Jacobsen, 2020).

Thirdly, ranking emerged as a central component of the way in which Facebook's throwback feature operated. According to Paluri and Aziz (2016), rankings were established and optimized to surface what was considered to be the 'right type' of memories, and so the delivery of automated memories produces certain types of engagements with the past. But, as Chapter One shows, these engagements with the past were engineered according to specific assumptions concerning what Facebook deemed meaningful and relevant. Paluri and Aziz (2016) state that they 'then rank potential memories so we can surface the most meaningful ones in the form of a News Feed story'. Here we see a distant echo of what Geoffrey Bowker (2008) called 'potential memory', the capture and organization of data for its potential usability in the future. As a result of processes of ranking, the automatic production of memory comes to be inextricably linked with the idea of *targeting*. As Paluri and Aziz (2016) point out, 'we then rank ... so we can surface'. The use of machine learning is framed here as the means by which this targeting is made possible and optimized, providing both 'better' and 'more accurate' predictions of what users will want to engage with. Yet it becomes increasingly evident that with features such as Facebook Memories, the conceptualization of users' past content as 'memories' comes to carry a range of connotations: memories are that which can be quantified, classified, ranked, predicted and, ultimately, targeted at users.

It is important to note that the ranking processes within the memories feature are facilitated by sustained attempts at inferring the level of meaningfulness of memories. Paluri and Aziz (2016) continue: 'The memories are ranked by a machine-learning model that we developed at Facebook. This model is trained in real-time and learns continuously; it gets better and more accurate at predicting what memories people want to see as they interact more with On This Day.' The inference of meaningful memories within the feature is framed as an iterative process. Meaningful memories are the desired outputs, and the ranking of memories is seen as a way to optimize and train the algorithm to reach that target. The aim is to predict the right type of memories, by which they mean the memories likely to suit social media's aims for continuous user activity. Moreover, the machine-learning algorithm developed at Facebook is depicted in highly promising terms, seeking to construct credibility in the way the throwback app functions. The depiction of machine learning is framed around a set of ideals, such as its coupling with real-time data analytics and the ability to learn continuously how to better anticipate what memories individuals will want to see. It is thus portrayed as a 'cluster of promises' (MacKenzie, 2013: 402), one of which contains multiple, overlapping selling points and seductive qualities. In a sense, the report by Paluri and Aziz represents Facebook Memories as being able to embody and deliver on the promises made by Big Data and data analytics (discussed in Beer, 2019). The notion of meaningfulness that emerges within this system is one that is based on a particular type of memory making and memory prioritization that is informed by the underlying logic and agenda of the social media platform.

The personalization and timing of the past

According to Paluri and Aziz (2016), the data being used to train this machine-learning algorithm falls under two overarching categories: personalization and content understanding. In order to personalize the experience of the memory feature, the engineers used signals such as 'previous interactions with On This Day (eg shares, dismisses), their demographic information (eg age, country), and the attributes of the memory (e.g. content type, number of years ago)' (Paluri & Aziz, 2016). Here the data feedback loops draw on the use of the feature to then inform its design and its future usage. Seeking out an increasingly personalized experience of the feature meant that Facebook was also able to variously engineer the frequency by which individual users were presented with memories. As Paluri and Aziz (2016) state: 'If a person has shared many memories from On This Day in the past, we can dial up the number of memories we show them in News Feed in the future.' Conversely, 'If a person has dismissed many memories, then we reduce the number of On This Day stories they see in News Feed moving forward.' The rhythms of remembering themselves become part of this targeting. In this case, we see that not only are memories ranked in terms of their relevance, but Facebook attempts to shape the routines and habits of those using the memory feature. They seek to shape the conditions by which users encounter their past as well as the frequency by which they are targeted with memories. As such, the automatic production of memories becomes, at least in part, what Shintaro Miyazaki (2016) calls 'algorythmic', a concept developed to examine the temporality, rhythmicity and materiality of algorithmic systems and how they operate in the world. The management of the flow of memories is also part of the broader algorithmic feed of content within social media, or what Elinor Carmi (2020) has called 'rhythmedia', emphasizing the importance of timing on platforms. Through an analysis of Facebook's ranking processes, it is apparent that the automatic production and targeting of memories partly depends on the rhythms established by algorithms. The targeting of memories is subject to their rhythms, movements and patterns of frequency, and thus has the potential to shape people's routines and habits in relation to how the past is encountered on social media. These social media are aiming to cultivate habitual and frequent connections with the past and to make past content an active part of the present.

In terms of the second set of signals used to train the machine-learning algorithm, the focus moves away from the user and to the attributes of the 'memory' itself. Paluri and Aziz (2016) state that On This Day relied on a computer vision platform, developed inside Facebook's Applied Machine Learning organization. Using this computer vision platform allowed the engineers to better assess and understand what exactly is contained within a particular photo or video. They describe that the platform:

is built on top of a deep convolutional neural network trained on millions of examples and is capable of recognizing a wide variety of visual concepts. It can detect objects, scenes, actions, places of interest, whether a photo or video contains objectionable content and more. It also has the capability to learn new visual concepts within minutes and apply them to incoming photos and videos. (Paluri & Aziz, 2016)

Here, the machine-learning algorithm is presented as a tool that provides a more granulated and 'pixelated' (Amoore, 2009) view of what is contained within each memory. Facebook's specific deployment of convolutional neural networks (CNNs) highlights how the power of algorithms resides in their capacity to render objects, events and people recognizable and visible. In the book *Cloud Ethics: Algorithms and the Attributes of Ourselves and Others*, Louise Amoore (2020) explores the ethical and political implications of machine-learning algorithms on society. CNNs, she states, commonly used for facial recognition and image classification, are able to recognize the attributes of a desired output, such as a particular object in an image, based on millions of weighted probabilities that weight some data points more than others. As such, Amoore argues, they need to be considered political actors in the world, embodying a 'regime of recognition' in terms of both 'arbitrating recognizability and outputting a desired target that is actionable' (Amoore, 2020: 72).

The neural network described by Paluri and Aziz can be seen as a particular 'regime of recognition' (Amoore, 2020: 67). That is, through the optimization of content understanding, the algorithm is able to deconstruct memories, as it were, in order to identify visual concepts, components and patterns within each memory. The computer vision platform is also represented as enabling fast-learning algorithms, capable of adapting to new information in a speedy fashion. The neural network is capable, it is suggested, of identifying patterns and visual concepts within people's memories on Facebook, which in turn enables it to learn new visual concepts and apply them to incoming memories. Based on this data-driven and pixelated understanding of what is contained within people's memories, Facebook's engineering team then used the signals from the machine-learning algorithm to rank the content (Paluri & Aziz, 2016). Thus, through personalization and content understanding a multifaceted algorithmic model emerged. On the one hand, it was able to render Facebook's memory taxonomy actionable to the ranking algorithm. On the other hand, it enabled what was perceived to be 'more accurate' ranking of people's memories based on different types of training signals. Memories, in this case, are rendered subject to the outputting of 'a desired target that is actionable' (Amoore, 2020: 72).

Although algorithms may appear stable and fixed, it is important to note that their outputs can also be fragile, highlighting the contingency inherent in the automatic production of memory on social media (on contingency see Fazi, 2018). As Amoore (2020: 75) observes, a small adjustment in the weights of a neural network may fundamentally change the output. This could mean that small tweaks, in the case we are examining, could radically alter the memories produced. This broader issue was demonstrated by Su et al's (2019) research paper, titled 'One pixel attack for fooling deep neural networks', where the researchers showed how the output of a deep neural network was radically and easily altered by changing a single pixel in an image. In terms of how algorithms 'see' the world, Su et al's (2019) research demonstrates how a single pixel might prove the difference between a car and a dog, and so on. This contingency is seen by Amoore as fundamentally political, since the 'basis for decision and action' (Amoore, 2020: 75) depends on and is shaped by the particularity of the output signal. In the case of the memories produced by social media and features such as Facebook Memories, it is similarly crucial to acknowledge the contingency, variability and fragility intrinsic to their production. As with the single pixel, a small adjustment in the weights of the neural network described by Paluri and Aziz might mean the difference between how memorable or not an image is considered to be by the social media platform. As a result, the classification and ranking of memories on Facebook should not be seen as fixed entities, but rather as automatically produced as well as inherently contingent.

Ontological promotion and the memory as Edge

Yet, we still need to ask what the implications are of these processes of ranking for the way memories are understood and made visible on the platform. What are the implications of ranking memories for the condition of possibility of people's memory making in everyday life? In order to better understand the role of ranking in the automatic production of memory, we turn to Pierre Bourdieu's (1984) notion of 'ontological promotion'. In his book *Distinction: A Social Critique of the Judgement of Taste*, Pierre Bourdieu (1984) examined the way aesthetic tastes crystallize and come to be seen as divorced from the classed reality in which they necessarily emerge. According to Bourdieu (1984: 6), tastes take on a 'sacred nature' that legitimates and generalizes their value in wider society. As a result, aesthetic tastes come to be perceived as fixed, as providing a stable standard by which to judge cultural outputs. Yet, Bourdieu argues, the power of such aesthetic tastes is that they do not simply remain imaginary. Rather, Bourdieu draws on the conception of transubstantiation to indicate how aesthetic tastes congeal and take on material form. Bourdieu (1984: 6) argues that culture 'does indeed confer on objects, persons and situations it touches, a sort of ontological promotion akin to a transubstantiation'. Bourdieu suggests that as aesthetic practices are inculcated over a long period of time, they are eventually transformed and ontologically promoted into legitimate expressions of what is considered 'cultured' or 'good taste'. Through ontological promotion, these aesthetic practices become flesh and blood, material realities.

Bourdieu's notion of ontological promotion is helpful in the context of Facebook Memories, especially in relation to the effects of ranking on memory. The concept helps with understanding the way that ranking participates in the ontological promotion of certain data points to memories, suggesting that Facebook Memories has the capacity to fundamentally shape the conditions by which memories are understood and come to matter in the world. Indeed, we argue that the automatic production of memories on social media is not only underpinned by the partitioning of the memorable, but also *the promotion of the memorable*. This, in turn, highlights the performative potential of algorithmic rankings. They constitute the conditions by which memories can be used for targeting on social media. As Kennedy et al (2015: 1) argue: 'Datafication should not only be understood as the process of collecting and analysing data about Internet users, but also as feeding such data back to users, enabling them to orient themselves in the world.' The ranking processes undergirding the automatic production of memory similarly produce weighted orientations (see Esposito & Stark, 2019). As people's past data is computationally resurfaced on social media, as 'memories' are fed back to them and used for targeting, the encounter with the memorable creates particular orientations towards memory, highlighting the politics inherent in all circulations of data (previously discussed in Beer, 2013).

As certain aspects of a user's past content are ontologically promoted into 'memories', throwback features such as Facebook Memories actively generate novel kinds of encounters between the past, the user and the platform. Through algorithmic processes of classification and ranking, past data is revitalized and 'memories' emerge as ontological promotions: something dug out, something new, something real. In one sense, the automatic production of memory entails that people encounter fragments of their past patterns of participation on Facebook. But, on the other hand, they are also targeted with memories to be cherished; memories that aim to add to users' overall social and mental 'wellbeing'; memories that include specific events, family and friends and which are framed as 'time well-spent' on the platform; in short, memories that have been classified and ranked in order to maximize engagement and participation. We argue that the ontological promotion of past content into targeted memories, through processes of ranking, have the capacity and potential to shape the conditions by which people encounter and remember the past.

We have shown how Facebook's throwback feature draws on past data, classifying and ranking prior content in order to predict what memories users will want to see resurfaced. The suggestion is that the way data is conceptualized and ranked on social media has a performative effect on the data itself. As Rob Kitchin (2014: 19) suggests: 'How data are ontologically defined and delimited is not a neutral, technical process, but a normative, political, and ethical one that is often contested and has consequences for subsequent analysis, interpretation and action.' In order to further illustrate how data can be ontologically promoted, we turn to Taina Bucher's (2012b) discussion of Facebook's underlying operational logic, the EdgeRank algorithm, and the regime of visibility it produces. As Bucher (2012b) points out in the article 'Want to be on the top?', social media platforms are fundamentally based around the mediation, shaping and production of visibility. The EdgeRank algorithm helps shape people's parameters of visibility on the platform, which in turn alters both what content they encounter and what counts as meaningful content. The way in which Facebook establishes relevancy is through a predetermined filtering and selection mechanism, where anything that shows up on a user's News Feed is considered an 'Object' by the platform. Objects include content such as posts and images uploaded to the platform (Bucher, 2012b: 1167). Any interaction users have with these Objects – through liking, commenting, or sharing – comes to be considered by the platform as an 'Edge'. These Edges signify vast heterogenous networks of relations, interwoven connections between users, businesses, events, products, cultural outputs such as movies or songs, and so on.

Bucher (2012b: 1167) identifies three key components in determining the rank and value of an Edge: affinity, weight and time decay. A highly ranked Edge, for instance, needs to encompass a certain affinity between the viewing user and the user uploading the image or writing the post; it must have a specific 'weight', meaning that it must fulfil Facebook's criteria for what they consider important, meaningful and conducive to engagement; and, lastly, it must be 'fresh', that is, it must have been recently posted or uploaded. Clearly this final category is being reconfigured in the case of the memory, with freshness being substituted in some way for nostalgia. As Bucher shows, the ranking and subsequent visibility of an Edge is determined by the multiplication of these computational factors (see Bucher, 2012b: 1168 for the EdgeRank formula). This means that 'becoming visible on the News Feed, appearing in that semi-public space, depends on a set of inscribed assumptions on what constitutes relevant or newsworthy stories' (Bucher, 2012b: 1168). Given this framework, the EdgeRank algorithm necessarily highlights some Edges while downgrading others, depending on their multiplication score. Practically speaking, this means that the parameters of visibility of content on Facebook are based on predetermined factors such as the specificity of the content, the person posting the content, the relationship between the content creator and the viewer, how many friends either 'like', 'comment',or 'share' the post, and so on. As a result, people's view of themselves, of others, and the world is shaped by the predetermined computational criteria established by the platform.

From the perspective of the platform, conceptualizing users' participation on the platform in terms of Objects and Edges is an effective way in which data is rendered amenable and actionable to the EdgeRank algorithm. In the case of this book, we are dealing with neither Objects nor Edges, but rather a set of data or content that cannot readily be conceptualized in terms of either category. As we outlined in Chapter One, the 'memories' that are resurfaced on Facebook constitute Objects, in so far as they refer to status updates and uploaded images. Yet they also constitute Edges, in so far as they comprise networks of connections as well as content that has been calculated and ranked according to the main criteria of the EdgeRank algorithm. Memories constitute Edges on Facebook in an additional sense, since when they resurface on users' News Feed in the present they become subject to the same computational criteria for relevance and visibility as other Edges. However, they also constitute something in their own right. That is, the way in which memories are classified, ranked, targeted and made to matter in the present varies fundamentally from the way Edges are calculated. We are, in a sense, dealing with an ontologically distinct class of data because, as we saw in Chapter One, not all past data come to figure in Facebook Memories. In fact, as we have shown, slices of dormant data are rendered amenable to algorithmic classification and ranking, which generates novel modes of participation in the present, novel networks of relations, as well as new ways of understanding oneself and others. Along with Objects and Edges, one can therefore also speak of *Memories* as a particular notion signifying the discrete increments of users' past content that have been ontologically promoted through classification and ranking. Valuable data is therefore no longer just what users post (Objects) or how they react to what others post (Edges); there is also data on how people variously react to past content being surfaced (Memories), whether it is their own or that of others. A subsection of a user's past content undergoes an ontological promotion to become memories on the platform.

Conclusion

Through the ranking processes outlined by Konrad (2017) and Paluri and Aziz (2016), past data is thus fundamentally reconfigured as something to be remembered, celebrated and shared – a memory. Through processes of classification and ranking, Facebook is extending its powerful reach backwards in time, drawing not only on real-time data but also real-time engagements with a resurfaced and targeted past in order to predict and shape the future. In the processes of the ranking and promotion of memory we directly see how the digging and excavation is being done on our behalf, changing, potentially, the way that memory operates and how the very notion of memories is understood. Memories are swallowed up into the broader sweeping logic of rating and ranking. As a result, they do not emerge organically from the data archives of Facebook. Instead, they are algorithmically constructed or 'dug out' (Benjamin, 1999a) through processes of classification and ranking, enabling them to be targeted at users within the platform.

What we have shown in this chapter is that something as potentially intangible as memories are open to the logic of metricization and calculability on social media. They are pulled within the remit of 'datafication' (Mayer-Schönberger & Cukier, 2013; van Dijck & Poell, 2013). In so doing, memories are reduced to a computational problem to be solved and optimized. Through the processes of classification, memories are rendered knowable to the algorithm. Once this occurs, memories, like any other measured object, can be ranked for their relative value or worth, depending on how these are defined. The ranking algorithm underpinning Facebook Memories also accentuates the visibility of some memories while ignoring others that are not featured highly enough within their classification framework. As with classification, it is the ranking that then sorts through which memories to surface, enabling the platform to target users with past content deemed memorable. As a result, the automatic production of memory on social media constitutes the dual process of partitioning *and* promoting the memorable. This dual process informs a particular version of our own biographies. Through processes of classification and ranking, targeted memories become akin to what can be seen as apparatuses of mattering. These are apparatus or mechanisms as well as practices that decide what matters and how things come to matter in the world (Barad, 2007: 148). In the next chapter, we turn to the way these targeted memories are received and negotiated by users in everyday life. Attending to their reception helps further our understanding of the classificatory struggles and the tensions engendered by the automatic production of memories on social media. In short, we begin to reflect on how the classification and ranking of memories are *felt* in everyday life.

4

The Reception of Targeted Memories in Everyday Life: Classificatory Struggles and the Tensions of Remembering

The kind of automatic production and targeting of memories that we have described in the previous chapters is still relatively new. Yet it is already widespread and deeply embedded in how people relate to their past through social media content. As we have shown, processes of classification and ranking are central to how people encounter past social media content as memories. What this will mean for collective and individual memory will take some time to fully understand. However, in this chapter we would like to turn to a project that was recently completed by the first-named author in order to begin to think through and explore what these changes might mean, examining how people might come to respond and react to these packaged and targeted memories. The previous chapters showcase how the memorable is partitioned and promoted. In this chapter, we will reflect more directly on the reception of the classified and ranked memories with which users are presented. Given the scope of the issues, this is not a complete endeavour, but it begins to give glimpses into the variegated reception of automatically sorted memories that might then be pursued further. It will indicate the types of direction that memory making may be taking in the context of social media and mobile devices. In short, this chapter begins to explore something that is well-established but little understood as of yet. As discussed in Chapter One, we may know some of what happens when digital memories or mediated memories become integrated, but this particular chapter is about how people react to *targeted memories*. Partitioning and promoting the memorable through processes of classification and ranking assumes that the memory categories produced are fixed and distinct (Mackenzie, 2015). Yet, as we shall show, the processes of classification and ranking do not necessarily mean that memories fit neatly into those fixed grids of Facebook's taxonomy; nor are the reactions entirely in keeping with those imagined in the rhetorical ideals of the social media providers and coders. As this chapter shows, the reception of targeted memories in everyday life emphasizes the various nuances and tensions generated by the dual process of classification and ranking.

Getting at the tensions of remembering

In this chapter, we draw upon 26 in-depth, semi-structured interviews with memory app users and four focus groups with participants with varying degrees of familiarity with social media. These interviews and focus groups were conducted throughout 2019. The data was collected as part of a broader project exploring the effects of algorithmic systems on people's memory practices and remembrance of the past. In this chapter, we are focusing only on the insights that are most relevant to the classification and ranking processes within these automated systems. The rationale for deploying both qualitative interviews and focus groups was to provide a more comprehensive and nuanced insight into the intersections of algorithms and memory in everyday life across a wide range of different types of social media users. Whereas the focus groups engendered a better understanding of the implicit and passive ways in which people react to being targeted with memories on social media, the qualitative interviews provided a means to further explore the reflexive and active engagement with memory features by those who are more frequent users. The qualitative interviews were conducted with people using the memory app, Timehop, as a particular avenue into the research area. Similar to Facebook Memories, Timehop is an app specifically designed to mine and resurface past data as 'memories' to users in the present. With over 21 million daily users (Lomas, 2018), the app is also highly popular, providing further indication of the scale at which social media users are already using these media as memory devices. Timehop, as a particular memory or throwback feature, was also selected since it was assumed that its user base comprised people using the app actively, intentionally and voluntarily.

Potential participants were identified and recruited through Twitter. Using Twitter as a sampling frame, the first author made regular searches on the platform between January and March 2019 for tweets containing words such as 'Timehop' and 'Timehop memories'. The assumption was that these tweets indicated some form of active or intentional usage of the memory feature. Afterwards, potential participants were contacted directly on Twitter and invited to take part in a Skype interview about their use of the memory feature.[1] Most of the participants who were interviewed routinely visited the app as part of their continual engagement with their own data pasts. Moreover, most of the participants also drew on their experience of using other memory features such as Apple Memories, Facebook Memories, or Snapchat Memories. The interviews were then audio recorded, transcribed and coded thematically, according to categories such as 'practices', 'affects', 'memories', 'perceptions of the app' and so on. The interviews provided insights into the variegated usage of a multiplicity of memory features, which provided grounds for comparative analysis.

The underlying aim of the focus group interviews, on the other hand, was to provide insights into the more implicit uses of and encounters with throwback features. The rationale for conducting focus groups was also predicated on the multiplicity and heterogeneity of contemporary memory features. Whereas some of these features comprise standalone apps such as Timehop, other prominent ones such as Facebook Memories and Apple Memories are more tangential features of already established platforms. Albeit tangential, these throwback features are inextricably integrated into and embedded within wider platforms or digital infrastructures. The assumption was therefore that the heterogeneity of memory features would necessarily have differential effects on how people experience being targeted by mediated memories. One of the aims of the focus groups was to juxtapose and discuss multiple memory features as well as their salient affordances, differences and similarities.[2]

The findings and themes generated from the participants' responses provided insights into the various ways that people use memory features and how they respond and react to being targeted by memories.[3] Furthermore, the findings provided an avenue to explore and examine the practical effects of processes of classification and ranking on memory practices 'in the wild' (Seaver, 2017: 2). Keeping the focus on the mundane impact of classification and ranking, this chapter examines what targeted memories do as part of specific situations as well as the various ways that participants responded to and negotiated these memories. Discussing a range of features such as Timehop, Facebook Memories, Snapchat Memories and Apple Memories, these analyses provide a prism through which to critically reflect on how the partitioning and promotion of the memorable is negotiated by users in everyday life. Yet, as we have already stated, these analyses constitute by no means the complete array of issues that targeted memories give rise to. Instead, in this section we seek to provide jumping–off points for further research needed in this area of ranking, classification and the reception of targeted memories. Firstly, we discuss the relationship between targeted memories and issues of attention.

"It directs my memories": technicity of attention

Being targeted by memories on various platforms and apps was considered, initially, by many of the participants to be a useful functionality. They were often said to provide participants with snapshots of different parts of their past, often reminding them of past experiences they had forgotten as well as providing useful focal points for social media interactions. As a result, statements like these were commonly seen in the interview data:

'I like the idea of an app that forces me to remember things that I didn't immediately remember.' (Diana)

'It'll trigger some connections in my brain to make me remember some strange part of my life that I'd forgotten.' (Becky)

'It brings up the moments that you might not ever search for.' (Imogen)

'It's not a memory that would have popped into my head or anything.' (Charlotte)

'I think it's nice to be reminded and it does make it a bit more personal, a bit more intimate.' (Francis)

For these participants, throwback features resurface and bring to mind past memories that they had momentarily forgotten or not thought about. This in itself is an indication that these automated systems have some impact on what is remembered, bringing to mind aspects of the past that people may not think about. It also suggests that the processes described in Chapters Two and Three may have a direct influence on the outcomes of remembering, often triggering forgotten connections. The participants' comments on the impact of these systems on what is remembered and when raise the question of how these particular memories were selected to be resurfaced. Drawing on Bourdieu's (1984) notion of ontological promotion, Chapter Three demonstrated how past data points are rendered meaningful to users in the present through processes of ranking, with some memories weighted as more memorable than others. Presenting a clear reason as to why this automated memory making has moved so swiftly into everyday practice and how it has reached such widescale use, it was often seen as a useful feature. As one participant noted, one of the main appeals of features such as Facebook Memories is that they are able to put memories "right in front of you" (Jack). However, as we have shown, far from being neutral, this operation is based on underlying classification and ranking processes that decide which memories are to be used for targeting. The appeal of throwback features, as a result, is their seeming ability to "direct my memories specifically", one participant noted, providing her with "memories of things that I had completely forgotten" (Alice). Certain aspects of the past are rediscovered or dug up on their behalf, they suggest. It is interesting to note that this is a practice that was frequently welcomed by respondents.

This capacity to direct targeted memories suggests that users' parameters of visibility and attention are inextricably interwoven with the underlying processes of memory features. As such, there is scope to understand targeted memories as a mode of what Taina Bucher (2012a) calls the 'technicity of attention'. Technicity of attention, for Bucher (2012a: 1–2), refers to the technical rationalities that are involved in the governing of participation on social media platforms (2012a: 1–2). Attention figures here as a crucial factor by which platforms seek to govern participation and increase engagement through the resurfacing of content considered worthy of users' attention. Attention is understood, in this context, as the mental or cognitive capacity to focus on a particular object or task at hand. Yet, instead of focusing on the definitional issues of attention, Bucher explores how attention is mediated and shaped by social media platforms. In the case of Facebook, its infrastructure is moulded around the capturing and processing of users' attention – something that memories appear to be achieving for these participants. In other words, the harnessing of attention is part of the infrastructural logic of the platform. This is done in order to instantiate modes of participation amenable to what has been called the 'like economy' (Gerlitz & Helmond, 2013). As Gerlitz and Helmond (2013) observe, the 'like' button becomes emblematic of the way in which platforms and third-party companies seek to know users on an increasingly granular level for commercial purposes, perpetuating what has been called 'knowing capitalism' (Thrift, 2005). Social media platforms then seek, as with the memories functions, to put something in front of a user that may motivate them in response to choose to press the 'like' button. For Bucher (2012a: 13), the 'like' button constitutes one mode by which Facebook's EdgeRank algorithm learns to select and rank information, calculating what is deserving of users' attention (the use of emojis may be seen as a further elaboration of this model). As such, a crucial way in which users' parameters of attention are shaped is through the particular content the platform deems deserving of users' attention. The notion of technicity of attention, therefore, highlights how platforms have the capacity to 'produce and instantiate modes of attention, specific to the environment in which it operates' (Bucher, 2012a: 4). Attention becomes encoded as information, engendering networks and connections between users as well as users and content, help to further produce and shape people's parameters of attention within the platform.

As our findings demonstrate, targeted memories become an equally crucial means by which platforms and memory features can produce modes of attention, specifically in relation to users' *data past*. This is particularly significant since some of the participants acknowledged the malleability of their memories and the potential for throwback features to shape them. As one of the participants stated, when asked about his reactions to seeing resurfaced memories in the present, "you're going to regurgitate it in a way that it's similar but it's not the same" (Ethan). This participant suggests that when memories are algorithmically resurfaced and targeted at users, they may generate novel affective states. In shaping people's parameters of attention, targeted memories become akin to what N. Katherine Hayles (2012) calls 'focusing actions'. For Hayles (2012: 14), 'focusing actions' are dynamic processes by which material realities are co-produced by humans and non-humans, perpetually changing 'as the focus of attention shifts'. The constant delivery of memories is perhaps one such focal point for action.

Equally, the encounter with targeted memories highlights memory making not as a static activity, but rather an imaginative negotiation and reconstruction of the past in the present. Of course, this notion of memory as an iterative and dynamic process is well established in the memory studies literature (see for instance, Hoskins, 2018; Kuhn, 1995; van Dijck, 2007). As Frederic Bartlett notably (1932: 213) put it: 'Remembering is not the re-excitation of innumerable fixed, lifeless and fragmentary traces. It is an imaginative reconstruction, or construction, built out of the relation of our attitude towards a whole active mass of organised past reactions or experience.' In a similar way, being targeted by memories on social media is not a passive experience. It initiates an interactive and iterative process of interpretation and reinterpretation, of 'imaginative reconstruction' of the past and our relationship to that past. Indeed, as Bucher (2012a: 9–10) argued, the notion of attention in relation to social media platforms should be seen as 'an emergent property of the interaction between technologies and users', as well as a property arising out of the 'software-subject continuum of Facebook' (2012a: 13). The targeting of memories through processes of classification and ranking should therefore not be understood as technologically deterministic. Instead, they engender a space where remembering can be co-instantiated, where the past is co-produced by the platform and users. This is where attention never remains fixed but always contingent on the interactions between users and the infrastructural logic of the platform.

This dynamic co-production of the past in the present also implies a complex and 'intimate entanglement' (Latimer & López Gómez, 2019) between what we understand to be memories in an embodied sense and the digital objects that mediate them. In one of the focus groups, an exchange between participants suggested that memories are made on the basis of the stimulus materials that are presented to us. In this account, the memories are made in relation to images of the past, suggesting that social media images classified and promoted as 'memories' may play a crucial part in what is retained and how the past is related to:

> Grace: Well, as I said, I think we can often create memory out of photographs. We may have no recollection at all of, say, a picture of yourself on a swing in the park when you were three or something. But, by the time you've seen that photograph a couple of times and sat with your mother and ooh you were wearing that dress and all this, it's true, it's in your head, isn't it. It's like you construct your memory from the photograph.
>
> Helen: I think the two become merged, don't they, it is difficult to dissociate one from the other.
>
> Grace: Yeah.

The classification of the image as a memory is important for how it is then responded to. In these moments, they acknowledge, it becomes hard to separate the image from the moment being remembered. These cannot easily be dissociated. In this view, social media images can become 'evocative objects' (Turkle, 2007), capturing emotions, affects, and aspects of our biographies. This seeming inseparability also echoes José van

Dijck's (2009) idea that mediated memories are an amalgamation of complex interactions between brain, embodiment, culture and technological systems. This would also suggest that there is scope for the classified and ranked memory to play a very direct part in shaping and defining what is remembered – that is to say, in shaping and partitioning the sensible, as Rancière (2010) put it (as discussed in Chapter Three).

The relationship between memories and the digital objects through which they are mediated creates potential problems for any dichotomies established between the organic and the inorganic, analogue and the digital, the embodied and the technological. As José van Dijck (2007: 28) argues: 'Mediated memories ... can be located neither strictly in the brain nor wholly outside in (material) culture but exist in both concurrently, for they are manifestations of a complex interaction between brain, material objects, and the cultural matrix from which they arise.' What we understand as targeted memories are not simply a prosthesis of the mind nor are they simply a repackaging of past data for effect. As we suggest in this book, the relationship between what social media platforms and users deem memorable are in a constant state of interaction, tension and constitutive mutuality. As a form of technicity of attention, targeted memories have the potential to direct and shape users' parameters of attention, bringing certain aspects of their data past to matter in the present, while overlooking others. In turn, people's engagement with these targeted memories and their perceptions of them help to mould the algorithm itself (see Bucher, 2017). For example, as we saw earlier in the book, how often users visit Facebook Memories determines the extent to which they will be targeted with past memories in the future. Distinctions between platforms, algorithms, cognition and embodiment are blurred as these are inextricably interdependent within a social media environment. The dual processes of classification and ranking signal that the conceptual and material parameters of what constitutes a 'memory', 'memory practices' and 'remembering' do not remain fixed. Rather, these are in a constant state of flux on social media, particularly as they are shaped and co-produced in conjunction with users and their data. Memories here become part of the broader tension around attention that is typical of social media spaces.

Further emphasizing the malleability of memory, one of the respondents noted, concerning her use of Timehop:

'I think it has altered the way I remember some things. Well, I was going to say, remembering it more objectively in the sense that I know the exact dates that I said or did something, but even more subjectively in the sense that I described just earlier. I might have embellished the point or I might have exaggerated and I don't know if what I said back then is that true in comparison to how I remember it. I do adjust my memories accordingly, I guess, in a way. You know, memories are very flexible.' (Taylor)

This notion of flexibility is significant because, as Louise Amoore (2009: 22) argues, the social power and politics of algorithms reside in the way they 'precisely function as a means of directing and disciplining attention'. Through processes of algorithmic classification and ranking, throwback features shape attention by calculating and predicting the types of targeted memories that are deemed deserving of users' attention. It follows that the idea of the past and what counts as meaningful memories may be shaped by the targeting of memories and by the act of directing users' attention towards specific data points, experiences or events in the past, ontologically promoting these to memories. In a sense, targeted memories, as a kind of technicity of attention, participate in the production of the past as much as they help remind people of it, capitalizing on the malleability and flexibility of memory.

"The algorithm doesn't really work with me": reductive algorithms

Another salient aspect of the way in which targeted memories are received in everyday life is in situations where they seem reductive to the recipient. Here, categorizations seem to have had the effect of being too limiting on the memory. In one of the focus groups, participants discussed their particular views of Facebook Memories and the way memories are classified, ranked, selected and targeted. One of the members stated that "I think it's nice to have Facebook remind us of one or two pictures, but the selection of pictures isn't always meaningful to me. Sometimes it's just some random photo" (Theo). This was echoed by Eva, who suggested that Facebook Memories "doesn't really correlate correctly, because I don't post any pictures. The algorithm doesn't really work with me. I think it's a cool idea, but it doesn't work if you don't post pictures". As well as highlighting a disjuncture, this point also emphasizes the visual aspects of these targeted memories. This led other participants to note how the frequency of general use of social media limited the feature's ability to locate or produce memories that might be considered rich or detailed: "Facebook I barely use so the memories are, if there are any, they are a bit irrelevant" (Olivia). The efficacy of memory features, it was observed, might rely more "on a frequent user of Facebook" (Elijah) and that "it definitely will work better the more that you use it" (William). The memory function, then, is contingent on the frequency of use and the type of content recreated in the past – otherwise the memories produced are seen as a little thin or lacking in meaning.

In these instances, the targeting of memories appeared ineffectual to users. Participants imagined that there might be several reasons for this. In Chapters Two and Three, we sought to highlight how the automatic production of memories on social media is based on the dual process of partitioning the memorable (classification) *and* promoting the memorable (ranking). In short, the targeting of memories is made possible by underlying processes of classification and ranking. However, when these memories seem to be based on insufficient data about the user, they appeared 'a bit irrelevant', resurfacing only 'some random photo'. The memory-making features seem to be based upon the frequent use of the platform and the logic of continual engagement. These cases highlight the interdependence of algorithms and data structures, considered by Lev Manovich (2001: 223) as 'two halves of the ontology of the world according to a computer'. As Manovich (2001: 223) writes in *The Language of New Media*, according to the logic of the computer:

> The world is reduced to two kinds of software objects that are complementary to each other – data structures and algorithms. Any process or task is reduced to an algorithm, a final sequence of simple operations that a computer can execute to accomplish a given task. And any object in the world – be it the population of a city, or the weather over the course of a century, or a chair, or a human brain – is modeled as a data structure, that is, data organized in a particular way for efficient search and retrieval … The computerization of culture involves the projection of these two fundamental parts of computer software.

As we have shown, targeted memories provide interesting insights into the specific operations performed on data by social media algorithms. Yet situations or instances where the algorithm appears reductive in everyday life foregrounds the dependence of targeted memories on a particular data structure, a model based on a specific understanding of memory and how it operates. Here we see that algorithmic outputs are not necessarily associated with people's actual experiences or understanding of the past. In other words, the targeting of memories is only as accurate as the data on which the ranking algorithm operates – it seems that the users are actually aware of this shortcoming based on their experience of the memories being promoted to them.

Another way in which targeted memories appeared reductive to the social media users was in the way memory features actively seek to predict which memories are worthy of people's attention. As some participants noted, this may engender certain tensions as well as feelings of unease. As one of the participants put it:

> 'If it's like Apple or Facebook or Snapchat they're picking this is probably what she wants to remember ... When I see it, I'm not like this is a photo album; I'm like this is a Snapchat photo album or this is what Apple chose. I'm never totally immersed in it. It's always through a technological lens ... you're always totally aware that they chose it.' (Eva)

The interviewee describes here how the automated selection of photos jars a little and acts to prevent them from becoming immersed in those "memories". The sorting of the content is just slightly out of place. The issue in this instance is not with the specific memories being used for targeting, but rather with the processes underlying their resurfacing. The algorithmic processes by which the memorable is partitioned and promoted is too pronounced, preventing the participant from being totally immersed in the version of the past that these targeted memories seek to draw attention to. The classification and ranking processes become, in a sense, too visible in the purposely designed and neatly packaged memory. Instead of directing people's attention towards the memory itself, the algorithmic systems instead draw attention to themselves, to their platformed nature and to their automatic production processes. It is well known that as algorithms are embedded within wider infrastructures in subtle and invisible ways, they often become visible in instances when the system crashes or if it performs in ways that users perceive to be incorrect (a point raised previously by Kitchin & Dodge, 2011). As we see here, the processes of classification and ranking seem too prominent to be missed by the recipient. As Eva put it, the underlying processes enabling the targeting of memories can be seen to engender a feeling of artificiality, of emotional detachment – 'a technological lens' where one is always aware of the choices that the platform makes. Returning to Walter Benjamin's fragment with which we opened Chapter One, this is suggestive of how the authenticity of the memory may be challenged when the individual has not dug it up or marked it out for themselves. This was also echoed by another participant who, when seeking to describe the feeling of encountering memories generated by the feature Apple Memories, stated, capturing the robotic and mechanistic feel of the interaction: "Hey, here is your trip, we built a memory for you!" (Quentin).

This notion of targeted memories as feeling somewhat artificial – as feeling like a *built* product – is particularly relevant considering a salient characteristic of algorithms more generally. In her work on the politics of algorithmic systems, Louise Amoore (2019, 2020) points out that for algorithms to work in particular contexts, their desired targets must be rendered manageable and actionable by reducing or condensing a multiplicity of possible outcomes to a single output. The danger of this, Amoore states, is that 'it is precisely the arbitrary groupings and attributes of the algorithm that become effaced when it becomes a technoscience for resolving political difficulty' (2019: 7–8). Through this condensation of multiple possible outcomes into one actionable output, algorithmic processes eradicate the contingency and arbitrariness that is inherent in how they function. This raises questions regarding the ways algorithmic systems correlate data points and generate relevant outputs. Luciana Parisi (2019: 4) suggests that 'as the system gathers and classifies data, learning algorithms therefore match-make, select and reduce choices by automatically deciding the most plausible of data correlations'. As Parisi points out, the condensation of multiple potential outputs into a single actionable one is a process marked not so much by certainty as by plausibility. This means that algorithmic predictions do not need to be 'exhaustive' as long as they are 'sufficient approximations' (Gillespie, 2014: 174). What we are seeing in the case of memories is that these approximations border on being insufficient in terms of the smoothness with which they circulate back into everyday life.

Targeted memories are an example of an outlet in which this algorithmic reduction of multiplicity is *felt* in everyday life. Echoing Amoore's observation on reduction or condensation, one participant described these targeted memories as having a "compressed" feel (Raymond), where the continuous past is rendered into discrete data points. The flow of life solidifies in ways that can miss the point or that simply feel too reduced down to be complete. As memory features seek to increasingly infer and rank 'memories' that are deemed worthy of users' attention, the condensation of the multiplicity and heterogeneity of a user's past content into one surfaceable memory may engender a sense of algorithms being reductive. This sense of the surfaced memory being reductive is supplemented by a sense that the algorithm, as one interviewee put it, "doesn't really work for me" (Eva). By stripping content out of a biography, targeted memories may therefore sometimes lack the necessary contextual information or metadata to appear meaningful to users, engendering an emotional detachment to the memorable item being produced by social media.

Yet, for some, this condensation or compression was conceived in less negative terms, suggesting that the reductive algorithm can produce more ambivalent outcomes. As Taylor stated, on her use of Timehop and Facebook Memories, "I was fascinated by just being able to collect all those pieces that you've shared online for so many years ... condensed in one spot". In this case, the algorithmic reduction of multiplicity helps to make the plethora of past data points manageable in the present – a perspective that fits with broader ideals about automated data analysis encapsulated in a kind of 'data imaginary' (see Beer, 2019). As another respondent observed, "It has helped me maintain riding shotgun on my inventory, so to speak, which is vast" (Anna). Although the condensation of multiple data points into one actionable output risks obfuscating the complexity of the past, it was also seen by some to provide a sense of agency as well as a means by which to navigate social media's 'vast unbounded sea of data' (Featherstone, 2000: 166) and draw meaningful memories from it. Through the algorithmic condensation of multiplicity, the data past was rendered tractable in the present.

Over time, as memories are repeatedly targeted at users, the algorithmic reduction of multiplicity can have a variety of effects. For some of the respondents, this recurring targeting of memories helped to cement a certain view of the past: "Every time you see that, it sears it into your memory even more" (Grace). Elsewhere it was reported that "What Timehop does it reminds you so constantly that I think that it's harder to forget things in the incorrect way" (Harvey). For other participants, the algorithmic reduction of multiplicity had a slightly different effect. When asked about her experience of being presented with memories, and if these had ever been the source of surprise, Diana responded:

'Absolutely. Although, as I have used the app more I found that because I have encountered myself of approximately 2007 to approximately, let's call the date in the past that I will the past, 2013, I feel like I have sufficiently watched that cycle. Some of the emotional edges have been worn off, and it's become a glossier past. I don't know. Having seen my yearly cycle of artifacts that I have created publicly in some ways I think it is worn smooth the messier or undocumented memories that I might have of that time.'

The algorithmic packaging of memories, this respondent suggests, creates a kind of glossy version of the past, one that the recipient themself feels uncomfortable or slightly disconcerted about. The biography is given a certain sheen when viewed through the logic of social media ranking. The question then becomes to what extent the user feels implicated in this process. As Diana stated later in the interview: "I don't want to be too Eternal Sunshiny about it. I don't want to have a hand in creating a rosy past. That feels artificial in ways that make me uncomfortable." These social media memories are quite obviously being packaged for them in ways that give the memory an artificial property; they have a sheen that does not sit comfortably with the individual's conception of their past. Here the reductiveness is caused by a rounding of the edges that leaves only a polished version behind. The algorithm appears reductive in its capacity to wear away emotional edges and smooth messier or undocumented memories. To this participant, it does not feel like a natural engagement with their own past, but seems more forced and superficial. Perhaps it is the lack of digging required by these pre-sorted memories that leads to these feelings toward the memories that are delivered. The memories do not arrive with enough residue of excavation upon them. As a mode of digging, the automatic production of memory reduces memories to the point of feeling artificial and therefore seemingly lacking in authenticity.

"When the algorithm goes wrong": algorithmic misconceptions

There were also occasions where the dual processes of classification and ranking of memories on features such as Facebook Memories resulted in clear misconceptions and mislabelling. As one participant stated: "I think it's a problem when the algorithm goes wrong and you still get those memories that you really don't want to see again" (Ava). As the interview and focus group data revealed, this feeling of the algorithm supposedly 'going wrong' was based on a few different factors. For Ava, for instance, the algorithmic misconception was felt through the targeting of memories she did not want to see resurfaced. Elsewhere, Diana identifies a similar experience of a misplaced targeted memory on social media, stating:

'There's something a little weird about [it], I've had all kinds of interesting anniversaries and things happen with this woman who was one of my best friends. Celebrating the first time that we hung out at a party feels like a strange thing to give too much weight to only because an app is showing it to me every year. I don't know, that's the thing I think about when I see that kind of cyclical stuff that I release to other people.'

In this case, the ranking processes appeared to give too much weight to a 'memory' that was not considered to be of particular significance to the respondent – plus the repeated rhythm also seems to place the wrong type of weighting on that content. Diana's statement seems to suggest that there is a crucial tension between what a platform is able to retain and what it should resurface – between what could potentially resurface and what should be remembered and shared. Classifying and ranking memories may therefore engender situations, such as in Diana's case, where users face 'memories' generally assumed to be memorable and meaningful, but which to the individual user may feel like "a strange thing to give too much weight to".

For others, such as Eva and Raymond, a sense of the algorithm getting it wrong was predicated more on the disjunction between the targeted memory on social media and what the respondents understood as a meaningful memory. The classification of what constitutes a memory was different for these individuals than it was for the platform. As Eva observed, discussing the Snapchat Memories feature:

'I do tend to write mostly. I write a lot. All of my "remember this?" there's nothing there to remember, which is strange to look back at. Because of the Snapchat

Memories, it's like "remember this?" … I'm like yeah, I guess I remember those things. They probably happened multiple times.'

There is a disjuncture here in this classification process around memory making. Similarly, Raymond said this about his use of Facebook Memories:

'I know that Facebook does it. I see the notifications, like it says you have memories with this person! I just don't understand it. A lot of these people are maybe just people that I know online or like met once or twice or like just knew but I don't really know in person anymore. I don't actually have things to be memories of. There's nothing to remember within. It's just that I sort of know them, I've talked to them a couple of times or whatever. So when Facebook says you have memories with – I don't know – John J. Smith, or whatever, whoever that is, I was like I guess so. I guess I added them to my Facebook and maybe interacted with them once or twice. There's no real memories associated with the memories that Facebook says that I have. Also, it seems to be like as with Facebook its algorithmic. It's trying to curate [inaudible] the like box, AI or whatever, but it's also trying to get me to use Facebook. That's the whole point of it is for me to use Facebook.'

As he faces content with, as he puts it, "nothing to remember within", Raymond's observation indicates some of the potential shortcomings of the dual process of partitioning the memorable and ontological promotion. There are mistakes or missteps in the predictions and reading of content that informs the ranking. The result is that which is made visible is sometimes ill fitting with the individual's understanding of their own past. This is illustrative of how a disconnect between Facebook's conception of a memory and the conception held by the individual can emerge. This represents a mismatch between the events and moments that the user considers to be worthwhile of the label memory and those that are being automatically categorized and then ranked as being of significance or meaningful by these systems.

More than mere mismatches, these cases may also demonstrate a kind of 'classificatory struggle' (Tyler, 2015) on Facebook. Where Imogen Tyler (2015) examines class relations as struggles against classification, here it is a struggle to define, partition and promote the memorable on social media. It is a struggle over how life and biography are classified, ranked and resurfaced. As Pierre Bourdieu (2018: 74) observes, the 'struggle over classification is a struggle for ... constitutive power, that is, the essentially political power to bring into being what is declared to exist'. The power of classification here is a constitutive power over the past. Nonetheless, despite this sense of a misplaced memory, this remains the version of biography that is presented back to the individual, shaping what is recalled and its outcomes.

As such, the struggle over classification in our context highlights the tensions between the 'memories' resurfaced through Facebook's ranking algorithm and what users remember about the past. These tensions can be understood in terms of what Kate Crawford (2016) describes as 'spaces of contestation'. That is, the spaces where algorithms and humans intersect are often marked by competing tensions, emotions, and passions. As Crawford (2016: 87) argues, 'algorithmic decision making is always a contest', that is they generate situations where humans and non-humans are often seen at odds with each other. For users such as Eva and Raymond, targeted memories may similarly engender affective tensions. What participants considered memorable could sometimes be seen in direct opposition to Facebook's attempt to partition and promote the memorable. This in turn highlights the struggles and tensions underlying the classification, ranking, and targeting of memories on social media as well as the various responses and processes of negotiations that arise from them. For participants like Raymond, their idea of a memory creates a problem for Facebook's promotion of the memorable, thus highlighting the classificatory struggles underlying algorithmic partitioning on Facebook. Of course, such instances also emphasize the active and creative memory work involved in all forms of engagement with and negotiation of the past (as discussed in Kuhn, 1995).

On one level, targeted memories and their underlying processes of classification and ranking can be seen to be prone to misconceptions of what constitutes a meaningful memory to users. Yet as we have also pointed out, there are many cases where the content labelled as 'memories' seems fitting, especially when it reminds users of past moments they have momentarily forgotten or where social media are algorithmically directing their memories in specific ways. In a crucial sense, the partition and the promotion of the memorable on social media has the capacity to shape the conditions by which memories are seen and understood as well as people's parameters of attention. The power of the automatic production of memories derives from the capacity to generate 'a whole variety of actuals' (Lash, 2007: 71) including a past increasingly relived through and within the algorithm.

On another level, these so-called algorithmic misconceptions reveal something further about the processes behind these systems as well as how they might come to redefine and reshape experiences and conceptions of memory in the future. As Taina Bucher (2018: 22) observes: 'What an algorithm signifies is an inherent assumption in all software design about order, sequence, and sorting.' This means that even when an algorithm supposedly fails in its intended task, 'it also speaks to us and tells us something of its limit points' (Amoore, 2019: 11). In this view, through these processes of classification and ranking, particular assumptions of the intersection of algorithms and memory come to the fore. The specific moments where the user's conception of a memory does not align with Facebook's provide glimpses into the inherent assumptions of these automated processes and their outcomes. In *Cloud Ethics*, Amoore (2020) explores particular instances where algorithmic systems are said to produce irrational outcomes – such as the case of 'the flash crash' of 2016, when the value of the British pound for some inexplicable reason fell 6.1 per cent in a matter of seconds (Amoore, 2020: 108). What Amoore (2020: 109) calls 'a kind of algorithmic madness, a frenzied departure from reason' should not necessarily be simply dismissed as an aberration which must be corrected and regulated. This would be to completely disregard the algorithm's dependence on a type of irrational logic. Instead, Amoore (2020: 110) argues that the line between reasonable and unreasonable actions is precisely 'the condition of possibility of algorithmic rationality'. Seen in this way, moments where algorithms supposedly act irrationally or incorrectly are valuable because they give us insight into the very form of rationality inherent in algorithmic systems. For Amoore (2020: 110), these are instances 'when algorithms give accounts of themselves'. Algorithms become more visible in their actions when they are seen to take missteps.

Targeted memories on social media can be understood in similar terms. Instead of exhibiting a misconception of what memories are, where the algorithm seemingly goes wrong in the view of the recipient, targeted memories provide insights into their underlying algorithmic processes as well as their fundamental assumptions about memory and the social world. As we pointed out in Chapters Two and Three, targeted memories foreground deep-seated assumptions about memories: that they can be datafied, classified and ranked, based partly on the attributes of Facebook's user population and what they deem memorable. Targeted memories are also conceptualized through a logic connected to Facebook's 'time well-spent' efforts, where past data likely to increase users' 'quality engagement' with the platform are partitioned and promoted into memories in the present (Facebook Newsroom, 2018; see Chapter One). Moreover, they are conceptualized through the positive emotional responses they may elicit, meaning that so-called 'bad memories' are filtered out and rendered un-resurfaceable (Jacobsen, 2020). As a result, the notion of algorithmic misconceptions suggests the potential tensions arising from targeted memories and their underlying processes of classification and ranking. Algorithmic misconceptions can be understood as instances where the algorithm gives account of its underlying classification and ranking processes. Yet, these misconceptions can also be seen as instances where algorithms give accounts of themselves, providing glimpses into the automated processes and their underlying assumptions.

"It leaves me a bit creeped out": invasive algorithms

Lastly, targeted memories may constitute a prism through which to explore some broader issues related to memory features and social media platforms in more depth. When asked what they thought the purpose was with memory features such as Facebook Memories and Apple Memories, one of the focus group respondents promptly answered: "Well, I think basically what it's doing is having access to all your past photographs." They added: "I would be very loathed to let third-party apps have access to my photographs" (Helen). This issue of access and data mining quickly developed into a question of being able to give permission, as is evident by the following exchange:

Helen: I think we all recognize the value of photographs as memories, don't we.

Anna: Oh yes!

Grace: Yes!

Helen: But we don't really want to sort of share those world-wide (laughter)

Grace: Well, we want to give the permission for that

Helen: Yes, yes indeed!

...

Grace: I find that one of the reasons I don't use Twitter and Facebook is that I find a lot of it to be a

> terrible intrusion of my privacy. It takes my
> time; it'll also use my photographs and things
> and I just don't like it. I don't want to live my
> life like that.

In this instance, targeted memories and the underlying processes of classification and ranking are seen to evoke anxieties related to some of the more entrenched issues of social media: data mining, lack of consent, invasion of privacy and so on. A member of a different focus group encapsulated these anxieties in direct terms, observing that "the fact that they [social media platforms] involve themselves and pick more stuff up about you leaves me a bit creeped out" (Lily). The idea of targeted memories, in these cases, is not so much a source of anxiety in and of itself. Instead, it is seen as emblematic of a wider trend on social media: the increasing mining of people's data, or what Couldry and Mejias (2019) have called 'data colonialism'. For Couldry and Mejias, data colonialism is both an external process, pushing the frontiers of what can be mined in terms of new areas of society, as well as an internal process, where increasing aspects of sociality that have not been previously mined fall under its purview. Understood in this way, the mining, classification, ranking and targeting of people's past data as memories is seen by some participants as yet another breach of boundaries of sorts or yet another way for social media platforms to 'pick more stuff up' about users.

In some cases, the notion of creepiness may foreground the rift between the norms of social media platforms and the norms of users more generally (Tene & Polonetsky, 2014). Yet, in the instances described in this chapter, it also highlights an ambivalence inherent in the interview and focus group data. Although some participants found targeted memories creepy and intrusive, others expressed a desire for an algorithm to provide them with 'more accurate' targeted memories (Imogen). This taps into ideals of ever greater personalization, ideals that are often in tension with senses of privacy. Yet the dream of accurate personalization persists alongside these discomforts. For example, one of the interview participants captured this ambivalence, stating: "It's like, oh my god it knows everything I did on March 1st. It's kind of freaky, but it's also kind of cool" (Keith). Other participants such as Imogen thought that features like Timehop would be more enjoyable if they developed an algorithm that made the targeted memories more accurate and predictive. This echoes the findings of Ruckenstein and Granroth (2019), where consumers were sometimes, on the surface at least, seen to want contradictory things: they oppose creepy advertising, yet wish for more relevant real-time analysis and predictions about the sorts of products they would like to buy. As such, the notion of being "a bit creeped out" encapsulates both some of the broader issues of social media as well as the ambivalences inherent in online targeting and how it is felt in everyday life.

Targeted memories were similarly conceived in ambivalent terms by other interview participants. Commenting on her use of Instagram and the way in which Timehop pulls up past Instagram posts, Miriam stated: "There's some things on Instagram, deep in my Instagram, that I haven't seen in a while too that I'm just like, oh I forgot that I posted this, like that's weird." This sense of discomfort, however slight it might be, derives from the participant wanting to have a sense of agency with regards to how memories are classified, ranked and targeted: "I like being in control of what is on my screen." At the same time, however, Miriam recognizes memory features as a source of comfort in her everyday life: "It's just comforting to know that I have something to keep all of my memories in order by year, by day." As such, the politics of targeted memories, and their underlying processes of classification and ranking, are fraught with tensions and ambivalences. Social media becoming memory devices creates this opportunity to not forget, but adds layers of targeting at the same time. A sense of creepiness does not necessarily signify users' antipathy towards targeted memories, but rather an ongoing complex negotiation of contradictory things: a resistance to their data being mined, but also an apparent and perhaps perceived desire to receive more accurate and relevant targeted memories (this competing tendency is discussed in Beer et al, 2019).

Another crucial issue that is raised when looking at the reception of targeted memories in everyday life is their long-term as well as socio-cultural effects. As has already been suggested, there is scope for classified and ranked memories to play a direct part in shaping and defining what is considered memorable, in shaping the sensible, as Ranciere put it (see Chapter Two). As a technicity of attention, targeted memories highlight the malleability of memory and how it can be directed and moulded in various ways. Yet, targeted memories are also oriented towards the future. As Yuk Hui (2017: 307) suggests: 'To remember something is always a reconstruction in which the fragmented past and the projected future are brought into the present.' Acknowledging this, one of the focus group participants pondered what the long-term effects of targeted memories might be:

> 'I don't know, you wonder if this ploy might give some sort of collateral damage with regards to your memory and your recollections of your friends and family over the years. I mean, it's only in its infancy now, isn't it? Twenty years on, when you've been reminded of a picture of your 21st birthday party however many times, what's it going to be like then?'
>
> (Grace)

Far from being oblivious, this suggests that users are seeing social media as archival technologies that intervene in the relations between the past and the future. The repetition of the annual reminder is also noted here. We find an unease with the automated systems deciding which moments to present back to users and how often they might do it. There is an acknowledgment of the power of the visibility and invisibility inherent in these systems and in the way that they partition and promote the memorable on social media. The power of the decisions of what type of memories should be encountered is important and, as we have begun to show in this chapter, mixes into what is remembered. Here, as Grace articulates in the excerpt, the social media user can become aware of how the system intervenes within and leads to memory production. While not being able to be sure about what impact this is having or how memories are surfacing in these systems and in their own consciousness, it is clear that some impact is occurring. Although this is beyond the remit of this book, there is scope to further explore the effects and socio-cultural implications of targeted memories and their underlying processes of classification and ranking of people's memory practices, as we will go on to discuss in Chapter Five. We need to ask, as Grace does, what is the potential 'collateral damage' of targeted memories on how people perceive and understand the past.

Conclusion

As this chapter has demonstrated, there are tensions that arise in conjunction with the automatic production of social media memories. These particular materializations of data–selves, to return to Lupton's (2020) conceptual framing (see Chapter One), are sometimes seen as potentially valuable while also producing a sense among recipients of inadequacy or that the self is not being correctly datafied. As we have made clear, this is an early attempt at trying to understand the sorting of social media memories and what happens once these sorted memories reinsert themselves back into the lives of individuals through these throwback features. We put these forward as a set of insights for future exploration. What we have found in this chapter is that as social media throws content back at individuals – following the classification and ranking of that content – it creates different types of encounters with the past. Moreover, it generates different kinds of reactions that all point toward the complex way in which content that is repackaged as memories will be received and reacted to. Rather than being comprehensive, we have picked out four features of these reactions in this chapter (we will discuss the possibility for a continuing research agenda in this field in Chapter Five). Firstly, there are the tensions around the directing of attention. Secondly, there are the tensions brought about when the sorting processes are seen to be too reductive. Thirdly, there are the tensions created by misconceptions of the targeted memory or of the content that is repackaged as such. And, finally, there are the tensions that arise when the algorithmic memory making is perceived to be overly invasive.

Back in Chapter One, we mentioned Andrea Mubi Brighenti's (2018) concept of 'measure-value environments' and suggested that we might see social media as exemplifying these types of environments in which measures and value interact. His point was that measures are not simply tools but are environments that we occupy. Within these environments, Brighenti suggests that we focus on the relations between measurement and measures on one side and value and values on the other (which relates to an earlier argument made by Adkins & Lury, 2012). Of course, these relations are not simply defined by a frictionless back and forth; instead, they are also sites of tension. To conclude this chapter, we want to briefly explore how the reception of targeted memories in everyday life, and the tensions outlined in this chapter, might relate to Brighenti's measure-value environments and the specific tensions that they generate.

Brighenti identifies 'three crucial tensions' that typify these measure-value environments. The first tension, Brighenti (2018: 30) claims, 'arises between *paradigm* and *syntagm*'. Brighenti explains:

> Measures are simultaneously formal standards and empirical practices: on the one hand, measures build upon a carefully defined, stabilized body of

knowledge, epitomized in the handbook of a certain scientific discipline; on the other hand, however, they are performed through practical – often even tacit – arrangements on the ground, which may thwart, or implicitly contradict, official procedures.

This first tension is between the formal version of knowledge and how that plays out in practice, with the realities pressing at the limits of knowledge. Categories may not quite fit or may not quite work in practice, and enumeration can cut out important properties or create frictions when hitting the contortions of the everyday. This first tension is thus between the form of classification as it hits the informality of practice (which echoes some of the tensions described by Bowker & Starr, 2000).

The second tension comes from the way that measures are used to know the world, which they then simultaneously also reshape. Brighenti explains that this 'second tension arises between *episteme* and *power*', adding that 'measures are ways of getting to know something about the world as well as, simultaneously, active tools to act upon the world and purposefully transform it' (Brighenti, 2018: 30; original emphases). As Espeland and Stevens (2008: 421) observe, metrics 'can become epistemic practices, embodying and routinizing norms of scepticism and certainty about the world'. The measures, as such, both capture and change the social world and how to act within it. For instance, Esposito and Stark (2019) argue that measures and rankings act upon the social world by producing orientations within it through visible reference points. By measuring we alter the social (as discussed further in Beer, 2016). This means that any attempt to understand the power of metrics and ranking processes also needs to consider that the very thing being captured is altered and potentially redefined as a result. By measuring the value of memories, for instance, we may also be changing those memories.

Finally, Brighenti (2018: 31) identifies a third tension in such measure-value environments. As he goes on to articulate:

> A third tension ensues, which concerns the unsettled relation between *means* and *ends*. Every time measures turn into targets they end up replacing the phenomenon they were supposed to apprehend in the first place ... Instead of measuring their current work, the production of measures and related preoccupations turns into an increasingly larger share of their work.

This third tension is about purpose. Brighenti argues that measures can easily become an end in themselves. Focus is therefore lost on the aims of the measurement and is replaced by the means. The result is a tension between means and end, in which the ends are challenged by the presence of the measure and start to blur with the means. The measurement takes on its own purpose and can even become the focus (at the expense of the thing being measured). Managing and feeding the measures becomes an end in itself. It even means that the measure can come to preoccupy attention and draw labour and interest away from the very thing being measured.

The four tensions that we have identified in this chapter do not map exactly onto the three tensions to which Brighenti is pointing, but we can see some shared properties that run across these perspectives. The types of tensions that Brighenti identifies in relation to knowledge, definition and purpose (if we can abbreviate them as such) are aimed at the wider 'social life of measures'. Yet we can still see them as in some way instructional in understanding the findings of this chapter. The three tensions that Brighenti identifies could be seen to underpin the four more specific tensions of the automatic production of memory that we have identified. The relations between measure and value are clearly associated with any attempt to decide for someone else, using a metric and ranking system, what memories they may wish to encounter. Brighenti's tensions are a useful framework for sensitizing us to the way that the tensions around the automated production of memory may be related to broader tensions that are brought about by the use of metrics and rankings to order and organize the everyday.

Essentially, when it comes to the classified and ranked memories of social media, this chapter has identified four tensions. These concern attention, reduction, misconception, and invasiveness. What this indicates is that we are seeing a tension emerge when these systems do not just deliver memories, but also actively seek to shape what is defined or understood to be a memory. A number of the points we have identified concern the way that the categorization of content creates friction with the definition of memory and classificatory understandings of social media users. As such, there is an additional level of classificatory work going on here, in which there are repeated attempts to reconceive the very concepts of memory and memories themselves. As José van Dijck (2010: 404) pointed out: 'As memories are increasingly mediated and thus constructed by networked technologies, the boundaries between present and past are no longer given, but they are the very stakes in debating what counts as memory.' As the content of memories becomes subject to automation, it has consequences for the very definition of memory. The algorithmic classification, ranking and targeting of memories can be understood as an intensification of the sort of boundary work that social media platforms do.

Finally, in closing this chapter, we return again to Walter Benjamin's fragment with which we opened this book. The alteration of the source of the authenticity of memory may be the source of the four tensions that we have identified. With the automation of the production of memory, it is not the toils of the individual that authenticate a memory; rather, it is the very notion of an automated, predictive and personalized system. Benjamin (1999a) noted how the individual act of digging and marking out of memories led to their perceived authenticity. In this chapter, we sought to show that when this digging and marking out are automated and targeted at users, then tensions of authenticity and value are likely to arise. Redefining what is a memory and where it comes from is always likely to be fraught with discomforts and disjunctures of different sorts. As well as giving insights into the responses of individuals to pre-packaged memories, we have tried to bring to the surface how these insights give glimpses into the very redefinition of memory as a concept. The reception of targeted memories in everyday life therefore provides novel insights into the potential tensions that arise when our biographies are ordered for us.

5

Conclusion: Sorting the Past

Even something as intimate and personal as memory cannot escape the reach of social media and their datafied and circulatory logic. In this book we have explored the underlying processes that enable the selection and targeting of past content in the form of repackaged 'memories'. Here we have highlighted the way that classification and ranking operate together to enable memories to resurface on social media throwback features. Through the combination of classification and ranking, the automated production and delivery of so-called 'memories' means that social media users do not need to dig; they are not excavating, as Walter Benjamin suggested, but instead that excavation is being done on their behalf. Benjamin noted that memories were always a way of mediating the masses of past experiences; this has not changed. These automated systems of social media remediate those memories through the classificatory systems that group them and then prioritize them, making them visible or invisible to us, and shaping how individuals and groups participate in those memories. Because, as Benjamin pointed out, memories have always been a mediation of the past, they can readily be reworked by these automated systems. As we have seen though, one problem with the automatic production of memory is authenticity. It is the act of producing memories that lends them authenticity; if that work becomes automated then potential tensions emerge around the legitimacy of that memory.

'The promise of automation', writes Mark Andrejevic (2020: 13), 'is to encode the social so that it can be offloaded onto machines.' In order to see the consequences this will have for memory and remembering, we suggest that there is a need to better understand the underlying classification and prioritization processes, what they are intended to do, as well as what implications and outcomes they have for people in everyday life. As a result, this book has sought to make a specific intervention into *the automatic production of memory*. Our contribution here has been to examine the role played by classification and ranking within these processes of automation. Once memories are opened up to classification and ranking, then the memories themselves will change, but so too will our understanding of what memories are. The concept of *memory* is unlikely to go untouched by these developments – indeed, we have sought to foreground the tensions that these processes of redefinition are already creating through features such as Facebook Memory.

In Chapter Two, we outlined how social media content is ordered within an existing classificatory grid. Focusing on a particular archetypal throwback feature, we showed how the 'Taxonomy of Memory Themes' was developed to classify people's content as 'memories' on the platform. Once boxed up, it is then decided where these bits of content are placed in terms of meaningfulness and positivity, and which of these 'memories' should become visible to individual users. Here we see the processes in which memory is actively being made within the archives of social media. Ronald E. Day's book *Indexing It All* (2014), explores how the kinds of classifications that accompany the escalation of content are now produced, and Facebook Memories serves as an apt illustration of this. As we show in Chapter Two, we are already seeing a widespread indexing of memory on social media platforms. We drew on Jacques Rancière's (2010) notions of the 'distribution of the sensible' and the 'partition of the sensible' to examine this development. For Rancière, politics constitutes the dual process of dispersing and spreading certain modes of perception and ways of understanding the social world, as well the process of dividing up the world, separating and excluding what counts as sensible and meaningful. In Chapter Two, we argued that social media are directly participating in the *partition of the memorable*, where processes of classification figure in the reconfiguration of past data into 'memories' while content not deemed memorable is sidelined. We argue that the partitioning of the memorable has the potential to shape the conditions by which past content is rendered meaningful and, in turn, shaping what can and should be remembered, celebrated and shared.

In Chapter Three, we turned to the processes of ranking that are a central part of the automatic production of memory. We showed how the taxonomization of people's memories provided not only the conditions by which to determine what counts as a memory and what does not, but also the conditions that enabled their ranking. This was the basis for how memories could be variously weighted and resurfaced within the throwback feature. As such, the processes of classification informed Facebook's ranking algorithm, which afforded the targeting of certain memories to certain users at certain times. Drawing on Pierre Bourdieu's (1984) notion of 'ontological promotion', the chapter argued that the automatic production of memories is not only underpinned by the partitioning of the memorable, but also *the promotion of the memorable*. In this framework, rankings participate in ontologically promoting users' past content into 'memories'. Clearly, the promotion of content into memories actively generates novel kinds of encounters between the past and present, between individual users, and between users and the platform. As a result, we argue that the ontological promotion of past content into targeted memories, through processes of ranking, has the capacity to shape the conditions by which people encounter and remember the past.

Put together, these focal points show how memory spaces are demarcated and how value judgments are then made in decisions about what are the right type of memories. These are now the means and mechanisms by which the past is encountered in everyday life. How these processes partition and then promote what is predicted to be a desired version of the past will have a powerful sway in the formation of biographies and in shaping *the view of life* that the automatic production of memory advances.

These processes, as we have noted, are not without their tensions. Just because something is packaged as a memory, does not necessarily make it so. These packaged memories may have implications and outcomes, even if they are not viewed in the same terms that has been intended. In Chapter Four, we turned to the reception of targeted memories, focusing on the ways in which these processes of classification and ranking come to be *felt* in everyday life. As that chapter highlighted, the categorization and prioritization of memory is not necessarily integrated into a smooth memory-making process. There are mismatches, strange choices and missteps. Drawing on interview and focus group data as well as Andrea Mubi Brighenti's (2018) notion of 'measure-value environments', we identified four points of tension generated in relation to the automatic production of memory on social media: attention, reductivity, misconception and invasiveness. Brighenti's tensions are a particularly useful framework for sensitizing us to the way that the tensions around the automated production of memory may be related to broader tensions that are brought about by the deployment of metrics and ranking to order and organize the social world. As such, the multifaceted reception of targeted memories in everyday life, we argue, provides novel insights into the potential tensions that are generated as a result of the partitioning and promotion of the memorable on social media.

What we have covered in this book is only one aspect of a broader phenomenon. We have focused here on the means by which so-called 'memories' are sorted in order to be targeted at the user. We focused on how past content is classified into a taxonomy of memory types and then ranked for utility in order to be targeted at individuals. Alongside this, we have offered some reflections on how people respond to the way that their 'memories' have been classified and ranked. On this latter point, we have only begun here to explore the responses that occur when people's 'memories', to return to Walter Benjamin, are dug or excavated on their behalf by automated systems. We have set this up as something that requires further examination in the future. Indeed, there are bigger questions that we hope this book will act as a foundation for. The repackaging of social media content as memories and its delivery to individual users is, we would argue, a fundamental aspect of social media. It is also, we contend, something that is likely to be reshaping individual and collective memory in as yet unknown ways. If social media are archives of everyday life, then they have profound possibilities for intervening in how memory is made and remade – the politics of the archive are just as pronounced here as in any archive (see Chapter Two). As social media become memory devices and as aspects of that become increasingly automated, we will need to rethink how memories work and how social media's redefinition and automatic production of memory might impact on processes of selfhood, person-making and collective interaction, among other things. By setting up the way that memories are classified and ranked on social media, we intend this book to act as a point of reference for understanding some of these broader and ongoing questions.

What is now needed is a sustained engagement with these broader questions and with the algorithmic interventions that are active in memory making, along with a careful exploration of what people make of these processes and how they respond to them. One key issue here is how the content that is defined as 'memories' in social media might intersect or potentially clash with the definitions and notions of memories that are held by social media users. The conception of what constitutes a memory becomes a potential site in which the implications of the categorization and ranking of memories can be opened up further. As a result, we have to consider what happens when classifications used by others and integrated into prominent media platforms may be seen as 'good' by those deploying them, while not necessarily leading to the prediction of behaviours and attitudes that it is imagined that they will provoke. This is indicative of a sort of classificatory struggle. As Pierre Bourdieu noted four decades ago, 'the good classification will allow us to construct the generative matrix from which we may predict behaviour and attitudes' (Bourdieu, 2018: 58). Given the generative power of classifications, it seems to us that understanding this needs attention, especially as social media are actively seeking to redefine and transform the means by which we approach and understand the past and our biographies.

Before closing, there is another area that is in need of further exploring. This is the role of the visual in contemporary memory making. Much of what we have dealt with here is visual in its form – or includes combinations of the textual and the visual. As this book has shown, memory features such as Facebook Memories rely heavily on the use of images and visuals in targeting users with memories from the past. The intersections of photography, perception and memory have already been much explored (see for instance Jurgenson, 2019, and, of course, the famous works by Benjamin, 2008; Berger, 2009; Sontag, 2008). In Chapter Four, we have begun to explore some of the effects of digital images on the way the past is remembered. This remained implicit in the chapter, and we would suggest that the visual dimensions of social media memory making require more direct attention. Given the malleability and flexibility of memory, Chapter Four suggested that digital images have the capacity to shift and shape people's perceptions as users' attention is directed towards specific data points, experiences or events in the past. Some participants, we pointed out, even found it difficult to distinguish between a memory of the past and the social media image facilitating that memory. As memory features such as Facebook Memories seem to rely on images for the algorithmic resurfacing of memories, the politics of the visual as well as the everyday practices of image capture and 'ubiquitous photography' (Hand, 2012) need to be factored into an understanding of the impact of social media on people's memories.

There is also a need for more critical thinking about the potential social and political ramifications of these developments for memory making on social media. One potential consequence may be the increasing *absence of voids* in future media landscapes. When it comes to social media, there is, to borrow a phrase from Jacques Rancière (2010: 36), 'no place for any void'. The repeated automated delivery of content relabelled as memories is one of the ways that social media seek to ensure that *there is no void* and that there is always something to respond to or engage with. The individual's past content provides a wealth of material for these platforms to actively draw upon in order to fill the void. The logic of these systems is that 'memories' are reduced to an assemblage of various past data points, aggregated and produced through a specific classificatory framework and ranking algorithms. In this view, the logic of partitioning and promoting the memorable is a way to ensure future engagement and participation on the platform. Memory is then redefined in these terms. Echoing Franco 'Bifo' Berardi's (2018) notion of breathlessness, there cannot be any space to breathe on social media platforms. These throwback features are fuel for this breathlessness. We suggest that the automatic production of memory may be emblematic of the continued efforts to displace any void on social media: constantly participate, constantly remember.

The concept of the *automatic production of memory* that we have pursued in this book is intended to open up possibilities for looking beyond the particular aspects that we have focused upon in this book. Indeed, the aim of this central concept is not solely to think of the combination of classification and ranking on which we have concentrated, but to incorporate an analysis of other archival and memory-making properties, to think through the implications of the algorithmic systematization of biography, to consider how people feel about their past being presented back to them in a selective way, to question how traces of the past circulate through networks, to examine the value and worth attached to past moments by these systems, to capture the rhythms of social media as they intersect with tides of life-cycles and, crucially, to think about how social media are actively reconfiguring the very notion of what a memory is – turning this concept towards a more narrowly defined version of itself that can then fit into the logic and limits of social media.

The past is now being sorted and targeted by social media; this is something that should not be overlooked. The changes that social media are bringing to the politics of remembering are too important to leave unchecked. Returning one final time to Walter Benjamin's fragment on memory, the juxtaposition of the messy memories depicted by Benjamin contrast sharply with the cleaned-up memories of social media. What is different in these two visions is the labour that occurs, and how important that labour is to the outcome. It is the very action of the *production* of memory that is at stake here. Based on digging and marking out, the production of memory that Benjamin described finds its authority, presence and meaning in the very act of production. What we have shown here is that when the marking out, in the form of classification, and the digging, in the form of ranking, become algorithmic the relations with those past events change, as do the type and form of memory that is encountered and the meaning attached to them. And so, because the processes of memory production are shifting, the automatic production of memory is about a transforming set of relations with the past that is continuing to unfold.

Notes

Chapter 1

[1]   This drive towards framing the Facebook Memories in terms of 'happy' memories or in terms of 'time well spent' is partly predicated on a high profile case from 2014. Eric Meyer, a web design consultant, wrote a widely publicized blog describing how Facebook had presented him with pictures of his recently deceased daughter on his suggested Year in Review video montage with the tagline 'Eric, here's what your year looked like'. The incident resulted in Facebook filtering out memories that they deem would have similar effects on other users (for more, see Jacobsen, 2020).

Chapter 4

[1]   The 26 participants who agreed to take part in the research project were then provided with an information sheet and consent form via email. The sample was demographically varied, both in terms of nationalities (for instance, US, UK, Canada, Costa Rica) and in terms of age (for instance, the sample ranged from 22 to 60). The interviews lasted, on average, between 30 minutes to one hour.

[2]   To procure a varied sample, purposive sampling was deployed, recruiting participants from both local community groups and students from the local university in an area of North Yorkshire, UK. These focus group members displayed various levels of familiarity with throwback features, ranging from those unfamiliar with these technologies to those using them or encountering them regularly, which resulted in varied and rich findings. Local community groups were contacted via email, and the university students were invited to take part in the project through departmental emailing lists. Although the focus group participants were recruited from the same area in North Yorkshire, the sample varied greatly in terms of age, ranging from 18 to late 70s. The focus group

lasted, on average, one hour. As a point of departure, the interviewer spent a few minutes at the start explaining how the features worked (ie Facebook Memories and Apple Memories), which the groups would be discussing. To aid the explanation, the interviewer also used visual aids such as screenshots of the interfaces of the relevant features. Although the discussions centred mostly on these two throwback features, participants also drew comparisons with other features such as Timehop and Snapchat Memories, which often provided interesting juxtapositions and points of contrast.

3   Both the qualitative interviews and the focus groups were audio recorded and transcribed, and the participants were pseudonymized. The interview and focus group data was coded manually and thematically, according to analytical categories such as 'practices', 'affects', 'memories', 'role of feature', 'perceptions of the feature' and so on.

References

Adkins, L. & Lury, C. (2012). Introduction: Special Measures. In Adkins, L. & Lury, C. (eds) *Measure and Value*. Oxford: Wiley-Blackwell. pp 5–23.

Ajana, B. (ed) (2018). *Metric Culture: Ontologies of Self-Tracking Practices*. Bingley: Emerald.

Ajana, B. & Beer, D. (2014). The biopolitics of biometrics: an interview with Btihaj Ajana. *Theory, Culture & Society*, 31(7–8): 329–36.

Amoore, L. (2009). Lines of sight: on the visualization of unknown futures. *Citizenship Studies*, 13(1): 17–30.

Amoore, L. (2019). Doubt and the algorithm: on the partial accounts of machine learning. *Theory, Culture & Society*, 36(6): 1–23.

Amoore, L. (2020). *Cloud Ethics: Algorithms and the Attributes of Ourselves and Others*. Durham, NC and London: Duke University Press.

Andrejevic, M. (2020). *Automated Media*. London: Routledge.

Barad, K. (2007). *Meeting the Universe Halfway: Quantum Physics and the Entanglement of Matter and Meaning*. Durham, NC and London: Duke University Press.

Bartlett, F. C. (1932). *Remembering: A Study in Experimental and Social Psychology*. Cambridge: Cambridge University Press.

Bartlett, J. (2018). *The People Vs Tech: How the Internet Is Killing Democracy (and How We Save It)*. London: Ebury Press.

Bechmann, A. & Bowker, G. C. (2019). Unsupervised by any other name: hidden layers of knowledge production in artificial intelligence on social media. *Big Data & Society*, 6(1): 1–11.

Beer, D. (2013). *Popular Culture and New Media: The Politics of Circulation*. Basingstoke: Palgrave Macmillan.

Beer, D. (2015). Productive measures: culture and measurement in the context of everyday neoliberalism. *Big Data & Society*, 2(1): 1–12.

Beer, D. (2016). *Metric Power*. Basingstoke: Palgrave Macmillan.

Beer, D. (2019). *The Data Gaze: Capitalism, Power, and Perception*. London: Sage.

Beer, D. (2020). Archive Fever Revisited: Algorithmic Archons and the Ordering of Social Media. In Lievrouw, L. A. & Loader, B. D. (eds) *Routledge Handbook of Digital Media and Communication*. London: Routledge.

Beer, D., Redden, J., Williamson, B. & Yuill, S. (2019). Landscape summary: online targeting: what is online targeting, what impact does it have, and how can we maximise benefits and minimise harms? 19 July 2019, *Centre for Data Ethics and Innovation*. Available at: https://assets.publishing.service.gov.uk/government/uploads/system/uploads/attachment_data/file/819057/Landscape_Summary_-_Online_Targeting.pdf (accessed 3 July 2020).

Benjamin, W. (1999a). *Walter Benjamin: Selected Writings, Volume 2, Part 2, 1931–1934*. Cambridge, MA: The Belknap Press of Harvard University Press.

Benjamin, W. (1999b). Theses on the Philosophy of History. In Arendt, H. (ed) *Illuminations*. London: Pimlico. pp 245–55.

Benjamin, W. (2006). *Berlin Childhood Around 1900*. Cambridge, MA: The Belknap Press of Harvard University Press.

Benjamin, W. (2008). *The Work of Art in the Age of Its Technological Reproducibility and Other Writings on Media*. Cambridge, MA and London: The Belknap Press of Harvard University Press.

Berardi, F. B. (2018). *Breathing: Chaos and Poetry*. South Pasadena, CA: Semiotext(e).

Berger, J. (2009). *About Looking*. London: Bloomsbury.

Bourdieu, P. (1984). *Distinction: A Social Critique of the Judgement of Taste* (trans Richard Nice). Cambridge, MA: Harvard University Press.

Bourdieu, P. (2018). *Classification Struggles: General Sociology, Volume 1. Lectures at the Collège de France (1981–1982)* (trans P. Collier). Cambridge: Polity Press.

Bowker, G. C. (1997). Lest we remember: organizational forgetting and the production of knowledge. *Management & Information Technologies*, 7(3): 113–38.

Bowker, G. C. (2008). *Memory Practices in the Sciences.* Cambridge, MA and London: The MIT Press.

Bowker, G. C. & Star, S. L. (2000). *Sorting Things Out: Classification and Its Consequences.* London and Cambridge, MA: The MIT Press.

Brighenti, A. M. (2018). The social life of measures: conceptualizing measure-value environments. *Theory, Culture & Society*, 35(1): 23–44.

Bucher, T. (2012a). A technicity of attention: how software 'makes sense'. *Culture Machine*, 13: 1–23.

Bucher, T. (2012b). Want to be on the top? Algorithmic power and the threat of invisibility on Facebook. *New Media and Society*, 14(7): 1164–80.

Bucher, T. (2017). The algorithmic imaginary: exploring the ordinary affects of Facebook algorithms. *Information, Communication & Society*, 20(1): 30–44.

Bucher, T. (2018). *If ... Then: Algorithmic Power and Politics.* Oxford: Oxford University Press.

Bucher, T. (2020). Nothing to disconnect from? Being singular plural in an age of machine learning. *Media, Culture & Society*, 42(4): 1–8.

Carman, A. (2018). Facebook is launching a new memories page to remind you of the days when Facebook was good. *The Verge.* Available at: www.theverge.com/2018/6/11/17442720/facebook-memories-on-this-day-month-launch (accessed 11 July 2019).

Carmi, E. (2020). Rhythmedia: a study of Facebook immune system. *Theory, Culture & Society*, 37(5): 119.

Constantine, J. (2018). Facebook feed change sacrifices time spent and news outlets for 'well-being'. *TechCrunch.* Available at: https://techcrunch.com/2018/01/11/facebook-time-well-spent/ (accessed 11 July 2019).

Couldry, N. & Mejias, U. A. (2019). Data colonialism: rethinking big data's relation to the contemporary subject. *Television & New Media*, 20(4): 336–49.

Crawford, K. (2016). Can an algorithm be agonistic? Ten scenes from life in calculated publics. *Science, Technology & Human Values*, 41(1): 77–92.

Davies, W. (2015). *The Happiness Industry: How the Government and Big Business Sold Us Well-Being*. London: Verso.

Day, R. E. (2014). *Indexing It All: The Subject in the Age of Documentation, Information, and Data*. Cambridge, MA: MIT Press.

Derrida, J. (1996). *Archive Fever: A Freudian Impression*. Chicago, IL and London: University of Chicago Press.

Espeland, W. N. & Stevens, M. (2008). A sociology of quantification. *European Journal of Sociology*, 49(3): 401–36.

Esposito, E. & Stark, D. (2019). What's observed in a rating? Rankings as orientation in the face of uncertainty. *Theory, Culture & Society*, 36(4): 3–26.

Eubanks, V. (2018). *Automating Inequality: How High-Tech Tools Profile, Police, and Punish the Poor*. New York: St. Martin's Press.

Experian (2018). Meet your data self. *Campaign*, 8 January 2018. Available at: www.campaignlive.co.uk/article/experian-meet-data-self-bbh-london/1453938 (accessed 3 July 2020).

Facebook Help Centre (2018). What things appear in Memories? Available at: https://en-gb.facebook.com/help/1422085768088554?helpref=uf_permalink (accessed 11 July 2019).

Facebook Newsroom (2018). All of your Facebook Memories are now in one place. Available at: https://newsroom.fb.com/news/2018/06/all-of-your-facebook-memories-are-now-in-one-place/ (accessed 11 July 2019).

Fazi, M. B. (2018). *Contingent Computation: Abstraction, Experience, and Indeterminacy in Computational Aesthetics*. London and New York: Rowman & Littlefield International.

Featherstone, M. (2000). Archiving cultures. *British Journal of Sociology*, 51(1): 161–84.

Foucault, M. (2002). *The Order of Things*. London: Routledge.

Foucault, M. (2008). *The Birth of Biopolitics: Lectures at the Collége de France 1978–1979*. Basingstoke: Palgrave Macmillan.

Garde-Hansen, J. (2011). *Media and Memory*. Edinburgh: Edinburgh University Press.

Garde-Hansen, J., Hoskins, A. & Reading, A. (2009). *Save As … Digital Memories*. Basingstoke: Palgrave Macmillan.

Gerlitz, C. & Helmond, A. (2013). The like economy: social buttons and the data-intensive web. *Information, Communication & Society*, 15(8): 1348–65.

Gillespie, T. (2010). The politics of 'platforms'. *New Media & Society*, 12(3): 347–64.

Gillespie, T. (2014). The Relevance of Algorithms. In Gillespie T., Boczkowski P. & Foot, K. (eds) *Media Technologies*. Cambridge, MA: The MIT Press. pp 167–93.

Hand, M. (2012) *Ubiquitous Photography*. Cambridge: Polity Press.

Hand, M. (2017). Persistent traces, potential memories: smartphones and the negotiation of visual, locative, and textual data in personal life. *Convergence: The International Journal of Research into New Media Technologies*, 22(3): 269–86.

Hayles, N. K. (2012). *How We Think: Digital Media and Contemporary Technogenesis*. Chicago, IL and London: The University of Chicago Press.

Hoskins, A. (2016). Memory ecologies. *Memory Studies*, 9(3): 348–57.

Hoskins, A. (2018). *Digital Media Studies: Media Pasts in Transitions*. New York and London: Routledge.

Hui, Y. (2017). On the Synthesis of Social Memories. In Blom, I., Lundemo, T. & Røssaak, E. (eds) *Memory in Motion: Archives, Technology and the Social*. Amsterdam: Amsterdam University Press. pp 307–25.

Humphreys, L. (2018). *The Qualified Self: Social Media and the Accounting of Everyday Life*. Cambridge, MA: The MIT Press.

Jacobsen, B. N. (2020). Sculpting digital voids: the politics of forgetting on Facebook. *Convergence: The International Journal of Research into New Media Technologies*. Online First, 1–14.

Jurgenson, N. (2019). *The Social Photo: On Photography and Social Media*. London and New York: Verso.

Keightley, E. & Pickering, M. (2014). Technologies of memory: practices of remembering in analogue and digital photography. *New Media & Society*, 16(4): 576–93.

Kennedy, H. (2016). *Post, Mine, Repeat: Social Media Data Mining Becomes Ordinary*. Basingstoke: Palgrave Macmillan.

Kennedy, H., Poell, T. & Van Dijck, J. (2015). Data and agency. *Big Data & Society*, 2(2): 1–7.

Kitchin, R. (2014). *The Data Revolution: Big Data, Open Data, Data Infrastructures & Their Consequences*. London: Sage.

Kitchin, R. & Dodge, M. (2011). *Code/Space: Software and Everyday Life*. Cambridge, MA and London: The MIT Press.

Konrad, A. (2017). Facebook memories: the research behind the products that connect you with your past. *Medium*. Available at: https://blog.prototypr.io/facebook-memories-the-research-behind-the-products-that-connect-you-with-your-past-f9a1d8a49a43 (accessed 11 July 2019).

Konrad, A., Isaacs, E. & Whittaker, S. (2016). Technology mediated memory: is technology altering our memories and interfering with well-being? *ACM Transactions on Computer-Human Interaction*, 23(4). DOI: 10.1145/2934667.

Koopman, C. (2019). *How We Became Our Data: A Genealogy of the Informational Person*. Chicago, IL: Chicago University Press.

Kuhn, A. (1995). *Family Secrets: Memory Acts and Imagination*. London and New York: Verso.

Kwek, D. H. B. & Seyfert, R. (2017). Affect matters: strolling through heterological ecologies. *Public Culture*, 30(1): 35–59.

Langley, P. & Leyshon, A. (2017). Platform capitalism: the intermediation and capitalisation of digital economic circulation. *Finance and Society*, 3(1): 11–31.

Lash, S. (2007). Power after hegemony: cultural studies in mutation? *Theory, Culture & Society*, 24(3): 55–78.

Latimer, J. & López Gómez, D. (2019). Intimate entanglements: affects, more-than-human intimacies and the politics of relations in science and technology. *The Sociological Review*, 67(2): 247–63.

Lawler, S. (2008). *Identity: Sociological Perspectives.* Cambridge: Polity Press.

Lomas, N. (2018). Timehop discloses July 4 data breach affecting 21 million. TechCrunch. Available at: https://techcrunch.com/2018/07/09/timehop-discloses-july-4-data-breach-affecting-21-million/ (accessed June 2020).

Lupton, D. (2020). *Data Selves: More-Than-Human Perspectives.* Cambridge: Polity.

MacDonald, R. L., Couldry, N. & Dickens, L. (2015). Digitization and materiality: researching community memory practice today. *The Sociological Review*, 63(1): 102–20.

Mackenzie, A. (2015). The production of prediction: what does machine learning want? *European Journal of Cultural Studies*, 18(4–5): 429–45.

Manovich, L. (2001). *The Language of New Media.* Cambridge, MA and London: The MIT Press.

Mayer-Schönberger, V. & Cukier, K. (2013). *Big Data: A Revolution That Will Transform How We Live, Work and Think.* London: John Murray.

Miller, C. (2019). *The Death of the Gods: The New Global Power Grab.* London: Windmill Books.

Miyazaki, S. (2016). Algorythmic Ecosystems: Neoliberal Couplings and Their Pathogenesis 1960–Present. In Seyfert, R. & Roberge, J. (eds) *Algorithmic Cultures: Essays on Meaning, Performance and New Technologies.* London and New York: Routledge. pp 128–39.

Moore, P. V. (2017). *The Quantified Self in Precarity: Work, Technology and What Counts.* London: Routledge.

Neiger, M., Meyers, O. & Zanberg, E. (2011). *On Media Memory: Collective Memory in a New Media Age.* Basingstoke: Palgrave Macmillan.

Noble, S. (2018). *Algorithms of Oppression: How Search Engines Reinforce Racism.* New York: New York University Press.

Özkul, D. & Humphreys, L. (2015). Record and remember: memory and meaning-making practices through mobile media. *Mobile Media & Communication*, 3(3): 351–65.

Paluri, M. & Aziz, O. (2016). Engineering for nostalgia: building a personalized 'On This Day' experience. *Facebook Research*. Available at: https://research.fb.com/engineering-for-nostalgia-building-a-personalized-on-this-day-experience/ (accessed 11 July 2019).

Parikka, J. (2018). The Underpinning Time: From Digital Memory to Network Microtemporality. In Hoskins, A. (ed) *Digital Memory Studies*. New York: Routledge. pp 156–72.

Parisi, L. (2019). Critical computation: digital automata and general artificial thinking. *Theory, Culture & Society*, 36(2): 89–121.

Prey, R. & Smit, R. (2019). From Personal to Personalized Memory: Social Media as Mnemotechnology. In Papacharissi, Z. (ed) *A Networked Self and Birth, Life, Death*. New York and London: Routledge. pp 209–24.

Rancière, J. (2004). *The Politics of Aesthetics: The Distribution of the Sensible* (trans Rockhill, G.). London and New York: Continuum.

Rancière, J. (2010). *Dissensus: On Politics and Aesthetics* (trans Corcoran, S.). London and New York: Continuum.

Reading, A. & Notley, T. (2018). Globital Memory Capital: Theorizing Digital Memory Economies. In Hoskins, D. (ed) *Digital Memory Studies: Media Pasts in Transitions*. New York and London: Routledge. pp 234–51.

Reid, S. (2019). Jogging happy memories. *American Psychological Association*. Available at: www.apa.org/monitor/2019/03/job-konrad (accessed July 2020).

Robinson, L., Cotten, S. R., Ono, H., Quan-Haase, A., Mesch, G., Chen, W., Schulz, J., Hale T. M. & Stern, M. J. (2015). Digital inequalities and why they matter. *Information, Communication & Society*, 18(5): 569–82.

Rottenburg, R., Merry, S. E., Park, S. J. & Mugler, J. (eds) (2015). *The World of Indicators: The Making of Governmental Knowledge Through Quantification*. Cambridge: Cambridge University Press.

Rouse, M. (2016). What is a data shadow? *WhatIs*. July 2016. Available at: https://whatis.techtarget.com/definition/data-shadow (accessed 3 July 2020).

Ruckenstein, M. & Granroth, J. (2019). Algorithms, advertising and the intimacy of surveillance. *Journal of Cultural Economy*, DOI: 10.1080/17530350.2019.1574866.

Savage, M. (2013). The 'social life of methods': a critical introduction. *Theory, Culture & Society*, 30(4): 3–21.

Seaver, N. (2017). Algorithms as culture: some tactics for the ethnography of algorithmic systems. *Big Data & Society*, 4(2): 1–12.

Sontag, S. (2008). *On Photography*. London: Penguin.

Srnicek, N. (2017). *Platform Capitalism*. Cambridge: Polity.

Su, J., Vargas, D. V. & Sakurai, K. (2019). One pixel attack for fooling deep neural networks. *IEEE Transactions on Evolutionary Computation*, 23(5): 828–41.

Tene, O. & Polonetsky, J. (2014). A Theory of creepy: technology, privacy, and the shifting social norms. *Yale Journal of Law of Technology*. Available at: https://digitalcommons.law.yale.edu/yjolt/vol16/iss1/2.

Thrift, N. (2005). *Knowing Capitalism*. London: Sage.

Thrift, N. & French, S. (2002). The automatic production of space. *Transactions of the British Institute of British Geographers*, 27(3): 309–35.

Turkle, S. (2007). *Evocative Objects: Things We Think With*. Cambridge, MA: MIT Press.

Tyler, I. (2015). Classificatory struggles: class, culture and inequality in neoliberal times. *Sociological Review*, 63(2): 493–511.

van Dijck, J. (2007). *Mediated Memories in the Digital Age*. Stanford, CA: Stanford University Press.

van Dijck, J. (2009). Mediated Memories as Amalgamations of Mind, Matter and Culture. In van de Vall, R. & Zwijnenberg, R. (eds) *The Body Within: Art, Medicine and Visualization*. Leiden and Boston, MA: Brill. pp 157–72.

van Dijck, J. (2010). Flickr and the culture of connectivity: sharing views, experiences, memories. *Memory Studies*, 4(4): 401–15.

van Dijck, J. & Poell, T. (2013). Understanding social media logic. *Media and Communication*, 11(2): 95–114.

Wu, A. X. & Taneja, H. (2020). Platform enclosure of human behavior and its measurement: using behavioral trace data against platform episteme. *New Media & Society*, online first.

Index